Undergraduate Lecture Notes in Physics

Series Editors

Neil Ashby, University of Colorado, Boulder, CO, USA

William Brantley, Furman University, Greenville, SC, USA

Matthew Deady, Physics Program, Bard College, Annandale-on-Hudson, NY, USA

Michael Fowler, University of Virginia, Charlottesville, VA, USA

Morten Hjorth-Jensen, University of Oslo, Oslo, Norway

Michael Inglis, SUNY Suffolk County Community College, Selden, NY, USA

Barry Luokkala (iD), Carnegie Mellon University, Pittsburgh, PA, USA

Undergraduate Lecture Notes in Physics (ULNP) publishes authoritative texts covering topics throughout pure and applied physics. Each title in the series is suitable as a basis for undergraduate instruction, typically containing practice problems, worked examples, chapter summaries, and suggestions for further reading.

ULNP titles must provide at least one of the following:

- An exceptionally clear and concise treatment of a standard undergraduate subject.
- A solid undergraduate-level introduction to a graduate, advanced, or non-standard subject.
- A novel perspective or an unusual approach to teaching a subject.

ULNP especially encourages new, original, and idiosyncratic approaches to physics teaching at the undergraduate level.

The purpose of ULNP is to provide intriguing, absorbing books that will continue to be the reader's preferred reference throughout their academic career.

Dhruba J. Biswas

A Beginner's Guide to Lasers and Their Applications, Part 1

Insights into Laser Science

 Springer

Dhruba J. Biswas
Former Head
Laser and Plasma Technology Division
Bhabha Atomic Research Centre
Mumbai, India

ISSN 2192-4791   ISSN 2192-4805   (electronic)
Undergraduate Lecture Notes in Physics
ISBN 978-3-031-24329-5   ISBN 978-3-031-24330-1   (eBook)
https://doi.org/10.1007/978-3-031-24330-1

This Springer imprint is published by the registered company Springer Nature Switzerland AG
The registered company address is: Gewerbestrasse 11, 6330 Cham, Switzerland

To my mother, GOURI,
And
To the memory of my late father, AJIT, whose influence on my life is too immense to express in words.

Preface

This book is intended to impart an enriching insight into the incredible world of an amazing source of light, the conception of which required a brilliant mind no less than that of Einstein himself when he envisaged the possibility of stimulated emission in 1917. That it took more than four decades for the experimental realization of **L**ight **A**mplification by **S**timulated **E**mission of **R**adiation speaks volumes about the complexity enveloping this fascinating concept. The first LASER flashing its incredible light in the year 1960 fired the imagination of scientists across the world. Dubbed in the beginning as a solution in search of problems, there is hardly any discipline left now where lasers have not made a cut. The fact that numerous Nobel Prizes to date have been awarded to brilliant pieces of work that involve lasers in one way or another bears testimony to its immense potential. Indeed, the invention of lasers by Maiman in 1960 has been a giant leap for humankind.

The beauty of lasers is not discerned by the majority, as most of the books pertaining to this grand invention generally cater to the needs of experts and researchers working in this field or graduate students aspiring to build a career in the field of optics or related areas. This book is aimed to satiate the thirst of that large cross-section of people, who have basic high school level physics background and inquisitive minds, but have not largely had the opportunity to appreciate the astounding science behind this extraordinary beam of light. Here, we attempt to elucidate the concept of lasers and many of their remarkable effects by invoking high school level physics and building on to it following an intuitive approach that avoids relying on complex mathematical jargon. A unique combination of the lucidity of text and clarity of illustrations without compromising the technical depth or scientific accuracy of the subject matter highlights our intent to capture the attention of not only the laser community but also a wide cross-section of undergraduate students, professionals from all branches of science, and amateurs. Furthermore, the text is often laced with anecdotes, picked from history, that are bound to pique the minds of the readers. Through this one-of-a-kind book, we intend to make the challenging physics palatable by refraining from using complex equations even as we enlighten our readers on: What is a laser? How does it work? What is its impact on humankind and, in particular, its influence on the advancement of science?

The three major aspects that would undoubtedly showcase a laser most comprehensively are the principle of its working, the role it plays in the advancement of science, and the influence it wields in our life. Consequently, the book has been planned in two parts to unmask all these three sides of the lasers. The endeavor of the first part is to bring to the fore the interesting science enveloping the operation of a laser. The next part is intended to be a comprehensive account of the impact of lasers on science and humanity.

My first encounter with a laser was over four decades ago when I witnessed an invisible beam of light emerging from a glass discharge tube to create a bright glow on a ceramic brick. I was awestruck and fell in love with this splendid source of light. As you read through this book, you would realize that this glass discharge tube was none other than a continuously working carbon dioxide laser, the most popular, versatile, and powerful laser in the infrared region of the electromagnetic spectrum. I shall be beyond fortunate if this book succeeds to kindle your curious mind.

Seattle, USA   Dhruba J. Biswas
November, 2022   (dhruba8biswas@gmail.com)

Acknowledgments

It gives me immense pleasure and joy to express my deep sense of gratitude to Padma Nilaya, my friend and colleague for the last three decades, without whom this book would not exist. She has been an endless source of academic support during the entire journey from the point of conceiving the idea to eventually turning that into this book.

I also would like to acknowledge the extraordinary debt I owe to Sam Harrison, Senior Editor, Springer Nature, New York. I have found him to be extremely helpful and always forthcoming in offering constructive and useful suggestions. He reviewed the book, chapter by chapter, and, I believe, he, through his excellent editing skill, has made this book eminently more readable than it otherwise would have been.

I am also indebted to Prasad Naik, my friend and former director of the Raja Ramanna Center of Advanced Technology, Indore, for his help in preparing the sections on semiconductor and free electron lasers. I am also profusely thankful to B. N. Upadhyaya, a young researcher of this center, for enlightening me on certain finer aspects of fiber laser physics and its working. I shall fail in my duties if I do not acknowledge the scientifically rewarding discussion with M. R. Shenoy, Emeritus Professor at the Indian Institute of Technology, Delhi, which I found to be very useful in explaining the subtleties of semiconductor lasers at an intuitive level. I also acknowledge with thanks many useful discussions and exchanges with my friend G. Ravindra Kumar, senior professor at the Tata Institute of Fundamental Research, Mumbai. I am also pleased to acknowledge the valuable scientific contribution of my colleague, Aniruddha Kumar, in the preparation of the manuscript.

I am also pleased to record heartfelt appreciation to my physicist friend, Prabir Pal, for his kind offer of reading through a substantial part of the manuscript. His observations and suggestions were very apt from a student's perspective. In this context, very useful comments from a few more friends, with non-laser backgrounds, are also thankfully acknowledged. They include Anup Gangopadhyay and Bapi Munsi, both physicists, Rajiv Karve, a chemist, and Satyabrata Sarkar, a chemical engineer. I am also thankful to Maninee Chaki, a young IT professional, and Souranil Ghosh, a young science student, for their help in the drawing of a few figures in Chaps. 2 and 7.

And, finally, a big thanks to my family, (wife, Modhumita, daughter, Debosmita, son, Debojyoti, and son in law, Ashis) for their constant support and encouragement, and, in particular, for granting me many long hours of working on the book. Ashis, a mechanical engineer whose research area has a distinct overlap with optical twee-zers, has also been instrumental in simplifying the seemingly intricate physics on multiple occasions in the text. I also acknowledge support and guidance of my octogenarian maternal uncle, Bishnu, a philanthropist, who has put his life to the service of underprivileged children of rural Bengal, in particular, by stimulating their young minds with scientific thoughts. And last, but not the least, words fail me to express love and appreciation to my little granddaughter, Coco, who as a toddler performed the role of a model to perfection to portray the real-world manifestation of the specular reflection of light (Fig. 2.6, Chap. 2).

Contents

Chapter 1
Introduction

The twentieth century dawned on a firm foundation of classical physics: courtesy of momentous contributions from Galileo Galilei (1564–1642), Isaac Newton (1643–1727), and James Clerk Maxwell (1831–1879), among many others. Galileo had the distinction of presenting a mechanical picture of the universe by connecting his prolonged experimental observation of celestial bodies with a theoretical description. Building primarily on the revolutionary work of Galileo and revealing observations on planetary motion by Johannes Kepler (1571–1630), Newton had put forward his laws of motion and gravity that together were able to fully predict the motion of a body, celestial or terrestrial. Nature, as we know, is made up of mass and (electromagnetic) radiation. Newtonian mechanics dealing with only the motion of a mass is, thus, just half of the story. This was complemented by the great advances made by Maxwell who, building on the conceptualizations of electric and magnetic fields primarily by Johann Carl Friedrich Gauss (1777–1855), Andre Marie Ampere (1775–1836), and Michael Faraday (1791–1867), developed electromagnetic field theory. The classical behavior of electromagnetic (EM) waves, such as interference or diffraction of light, can be completely described by Maxwell's equations. The EM wave theory, however, stumbled on a puzzling phenomenon called the photoelectric effect [1] brought afore in 1902 by Philipp Lenard (1862–1947), a staunch German nationalist and Nazi sympathizer. He discovered that the maximum energy of the electrons emitted from a metal surface upon shining by ultraviolet light is independent of the intensity of the light. EM wave theory is in complete contradiction to this observation, as according to this theory, the greater the intensity of shining light, the greater the energy and, in turn, the velocity of the electrons emitted by the metal surface. At this critical juncture, Albert Einstein (1879–1955) made his appearance in the year 1905, a momentous year in the history of science. He transformed physics in general, as he unified mass and

© The Author(s), under exclusive license to Springer Nature Switzerland AG 2023
D. J. Biswas, *A Beginner's Guide to Lasers and Their Applications, Part 1*,
Undergraduate Lecture Notes in Physics,
https://doi.org/10.1007/978-3-031-24330-1_1

Fig. 1.1 Sun, being astronomically apart, has its rays reaching the Earth almost parallel to each other. A sizeable concave mirror can catch a significant part of sunlight and focus it almost to a point (F). This can result in the generation of intense heat at the focal point

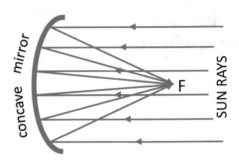

radiation, the two facets of nature, into $(E=Mc^2)$[1] through his theory of relativity, and optics, in particular, by conceptualizing the quantization of light. The realization that light can manifest both as wave and as particle rescued the vexed scientific community from the puzzle surrounding the photoelectric effect and placed the EM theory back on a firm footing. It is ironic though that the discoverer of the photoelectric effect, Lenard, a Nobel Laureate himself, termed Einstein's illuminating insight into this phenomenon, seemed in the beginning too challenging to be deciphered, as a misleading idea of "Jewish physics." Einstein's genius did not stop here. The enigma of light is known to have pervaded his mind ever since he was a child[2] [2]. He will soon put forward a revolutionary concept that would turn optical physics on its head planting the seed for a device, with no parallel in harnessing the power of light, that has intrigued the human mind from time immemorial.

Dating back to the medieval era, harnessing light has been a key to the evolution of civilization (The readers may refer to Chap. 2 dealing with the properties of light in this context). For example, legend has it that Archimedes (BC 287–212), the famous Greek scientist and mathematician, created a "death ray" by reflecting and then focusing a sizeable cross section of sunlight through a concave mirror of enormous size onto the Roman ships that were attempting to invade Greece (Fig. 1.1). This myth claims that Archimedes succeeded in burning down the Roman fleet at the siege of Syracuse in BC 213 sending the enemy army into complete disarray. It is of interest to note here that British scientists A. A. Mills and R. Clift investigated this historical legend and published a paper in 1992 entitled "Reflections of the Burning Mirrors of Archimedes" in the *European Journal of Physics* [3]. The study concluded that 440 men, each holding a mirror of area $1m^2$ or possibly a single giant mirror of appropriate size, may be able to just ignite a tiny area of a wooden boat placed at a distance of about 50 meters. There are, therefore, insurmountable difficulties in burning an entire ship using this technique.

---

[1] $E = Mc^2$ is regarded as the most famous equation in physics and perhaps in science. It underlines the equivalence of mass and energy and their interconvertibility. This seemingly simple looking elegant equation, however, has ramifications of colossal magnitude – it planted the seed for the atom bomb, the most devastating weapon produced by humans.

[2] By Einstein's own recollection, the nascent thought of invariance of the speed of light that would later give birth to the theory of relativity and change the world forever struck him at his early teen.

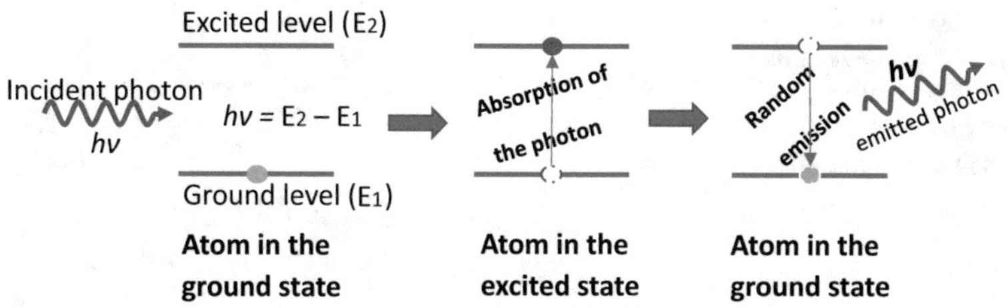

Excited level (E2)

Incident photon
$h\nu$ = E2 − E1

Absorption of
the photon

Random
emission

$h\nu$
emitted photon

Ground level (E1)

Atom in the
ground state

Atom in the
excited state

Atom in the
ground state

Fig. 1.2 The process of spontaneous emission has been schematically represented in the traces of this figure. An atom upon shining by light of appropriate frequency (left trace) absorbs a photon and gets electronically excited (middle trace). The atom can only reside in the excited state for a finite time following which it drops down to the ground state releasing the energy of excitation as another photon (right trace). The photon can be emitted randomly in any direction. In reality there will be an ensemble of atoms leading to emission of multiple photons moving in all possible directions as depicted in the next figure

Archimedes' soldiers perhaps used the mirrors to blind, confuse, or even burn the on-board Romans, the intensity (power per unit area) of sunlight reaching the Earth being too little to make the scheme viable as a matter of fact. Archimedes possibly wondered if only sunlight were a little more powerful! Just as the Sun does, all other conventional light sources too emit light all around them, mathematically speaking over the entire 4π solid angle[3] surrounding it. This sounds quite logical too! How on Earth would the source throw more light in a particular direction?

Let us make a small digression here and consider the case of a fluorescent source of light (refer to Chap. 4 to know the working of a fluorescent light) to understand this fact more clearly. According to the atomic model proposed by Niels Bohr (1885–1962) in 1913 [4], an electron of an atom can rotate around its nucleus only in certain specific orbits. As explained in Chap. 3, an electron can acquire energy only in a discrete manner, and the higher the energy of the electron, the longer its orbiting radius. Two such energy levels are shown in Fig. 1.2, the ground (E_1) and the excited (E_2) levels with the orbiting radius of the excited level being longer than that of the ground level. The ground state electron, as we know from Bohr's theory, can absorb photons from incident light if its energy ($h\nu$) matches the energy difference of these two levels (i.e., E_1–E_2) exactly. This electron then makes a quantum leap into the excited state, termed as absorption of photons (middle trace of Fig. 1.2). The electron is able to reside in the excited state only for a finite time and eventually falls back into the ground level by releasing its energy of excitation as a second photon, with as much energy as that of the incident photon, travelling in any random direction. This process is called spontaneous or random emission (right most trace of Fig. 1.2). Unmistakably, the incident light will comprise numerous photons,

---

[3] A solid angle is a three-dimensional analog of an angle. It can be readily understood by considering the cases of a circle and a sphere. A circle of radius r has a perimeter $2\pi r$ which subtends an angle 2π at the center, while a sphere of radius r has a surface area of $4\pi r^2$ which subtends a 3D angle or solid angle 4π at the center of the sphere.

Fig. 1.3 2D view of an ensemble of excited atoms releasing their energy randomly by emitting photons in every possible direction

Fig. 1.4 2D representation of a point source of light 'S' emitting in every direction. Energy emitted by it at a given instant will lie after a time 't' uniformly on the surface of a sphere of radius 'ct'

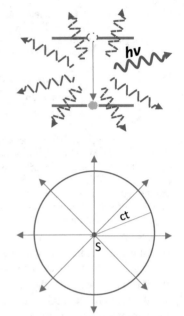

and there will be an ensemble of atoms. Thus, in reality, there will be an emission of multiple photons moving in all possible directions practically over the entire 4π solid angle around the atomic source. Figure 1.3 is a schematic 2D representation of the random emission of the photons. The same will also hold true for a thermal source of light that is more abundant in nature. To conclude, a conventional source of light will send light uniformly all around it.

In order to recognize the consequence of this fact, let us consider the case of a point source "S" of light as represented in Fig. 1.4. We assume that this source is capable of emitting "N" photons each of energy "ε" at a particular instant of time.[4] These photons will travel in all possible directions and, thus, after a time "t" would reside on the surface of a sphere of radius "r = c × t" (where c is the speed of light) maintaining a uniform density across it. It is straightforward to show that the energy "E" of the photons per unit area on the surface of this sphere will be given by

$$E = N\varepsilon / \left(4\pi r^2\right).$$

The energy emitted by a typical source of light over a given time per unit area is thus intrinsically low and continues to drop by the inverse square law with increasing distance. A 100 Watt traditional light source can therefore deliver 100/4π, i.e., ~8 Joules of energy per second over a 1 square meter area at a distance of 1 m, ~2 J at a

---

[4] A scientifically correct statement would, however, be that these N photons are being emitted by the source over an infinitesimally small duration of time.

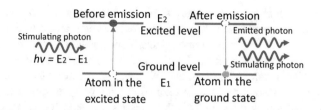

Fig. 1.5 A photon with an energy equalling the energy difference of the excited and ground levels, called the stimulating photon, upon interacting with an excited atom (left trace) can force the atom to drop down to the ground level releasing its energy of excitation as a second photon travelling in the same direction as the stimulating photon (right trace) and also matching it's all other attributes. Directionality is thus inbuilt in the process of stimulated emission

distance of 2 m, ~0.89 J at 3 m, ~0.5 J at 4 m, and so on. The random nature of emission of light obviously makes a light source appear progressively weaker with increasing distance.

As a departure from this well-known fact, Einstein, in 1917, theorized the possibility of a source beaming light in a particular direction through a mechanism called stimulated emission. To gain insight into this seemingly implausible process of emission of light, we once again focus our attention on the atom-photon picture we just considered. We begin here from an atom that has been electronically excited from its ground level (Fig. 1.5). Left to itself, this excited atom, as we know now, quickly falls back to its ground state releasing its energy of excitation as a photon in a random direction. Here Einstein's mind was intrigued by an ingenious thought – what happens if this excited atom comes in contact with a photon having energy exactly equal to the energy of excitation of the atom? This eventually led to the conceptualization of stimulated emission in a paper he published in 1917 [5], which laid the theoretical foundation of lasers, the light of twentieth century and beyond. In a nutshell, if this excited atom can be made to interact with a photon possessing exactly the energy of excitation of the atom, it will force the atom to *instantaneously* drop down to the ground level emitting a second photon that would replicate the first photon in all respects including its *direction of travel*. The occurrence of stimulated emission has been schematically represented in Fig. 1.5.

It is no wonder that a source of light, if its emissions can be forced through stimulation, would throw its entire light in a particular direction. Ideally speaking, its intensity will, therefore, not diminish with distance in complete contrast to all the other sources of light hitherto known to humankind. The bottom line is undoubtedly that the translation of the concept of stimulated emission into reality is contingent on how the countless emitting species in a macroscopic light source can be forced to emit in an orderly manner. Taking the cue from the way stimulated emission manifests, it would seem that the trick would be to bring a resonant photon[5] into an ensemble of excited atoms. This photon would then force the first excited atom it

---

[5] A resonant photon in this case is the one whose energy equals the energy of excitation of the atom.

interacts with to emit a second identical photon moving in unison with the first. These two photons would then stimulate two more excited atoms to emit two more identical photons making their number four, which would generate another four photons, and so on, triggering a rapid buildup of their number all moving in the same direction. While picturing the buildup of one photon into many this way, we understandably presumed that all the excited species wait for stimulation and that our photons cannot be lost through another process. Let us delve a little deeper into these two crucial issues. It should be remembered that although we have considered an ensemble of atomic species here, the underlying physics will remain unchanged for any other species such as molecule, ion, or whatever.

It is well known that nature prefers to maintain a state with the lowest energy, and if this equilibrium condition is disturbed by some means, it would find ways to return to its minimum energy condition. Under ambient conditions, all the atoms of our ensemble, therefore, reside in the electronic ground state that has the minimum energy. If by some means the atom has been excited to a level with higher energy (e.g., by shining it with light of appropriate wavelength; revisit Fig. 1.2 and related text in this context), it will come back to its ground state very rapidly, typically, say, within several microseconds or so. In other words, the lifetime[6] (t_L) of the atom in an excited state can be in the range of microseconds. This does not mean that once excited, the atom will always remain excited for this much time before returning to the ground state. The decay to the ground state is basically a random process, and the atom can spend any time between zero and t_L before returning to the ground state. An ensemble would comprise countless atoms, and if many of them are excited, some may decay to the ground state almost instantaneously (~0 time), while some would decay after spending the full lifetime (t_L), some returning somewhere in between, and so on. The situation is akin to the case of radioactive decay of an ensemble of unstable nuclei; some would decay just at the instant you are reading through this line, while some may decay after years! In contrast, a photon can stimulate an excited atom to emit instantaneously and can, therefore, easily beat the process of loss of excitation through random emission. Our first assumption that most of the excited species will be available for stimulation is, therefore, valid.

Let us now take a closer look at the interaction of the resonant photons with an ensemble of atoms. If the photon interacts with an excited atom, it can, of course, stimulate the atom to emit another identical photon. However, if it happens to interact instead with a ground state atom, the photon will be consumed as the atom will simply absorb it. Thus, the second assumption that our photons cannot be lost through another process is untrue. A stimulated emission adds one photon into the system, while the process of absorption removes one photon from the system. Thus, there can be a net gain in the number of photons if only the rate of stimulation can beat the rate of absorption. This understandably will be true only if the number of excited atoms exceeds that in the ground state. In the ensemble we began with,

---

[6] In the normal parlance, the lifetime of an excited energy state can be defined as the time over which ~66% of its population decays to the ground energy state.

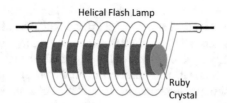

Fig. 1.6 This is a schematic representation of the experimental arrangement Maiman used to create the first laser. As a voltage pulse is impressed upon the helical flashlamp, it gives out light that shines on the ruby from all sides. The ruby, in turn, throws a stream of red light through one of its faces

however, all the atoms were in the ground state. It is, therefore, obvious that light amplification through stimulated emission will be possible only if more than half of the ground state atoms can be somehow lifted to the excited state, a requirement that once again seemingly violates the rule of nature (the readers are advised to take a look at the Boltzmann distribution described in Chap. 3). Although this is a major obstacle for achieving amplification of light through stimulated emission, you will recognize, as you read through the future chapters of the first part of this book, the presence of more hurdles that need to be overcome for the realization of this phenomenon. The fact that it took more than four decades for the experimental realization of light amplification by stimulated emission of radiation speaks volumes about the complexity enveloping this fascinating concept.

A laser was operated for the first time precisely on May 16, 1960, by an American physicist Theodore Maiman (1927–2007) at the Hughes Research Laboratory in Malibu, California. A small cylindrical synthetic ruby crystal, placed inside a spring-shaped flashlamp (Fig. 1.6), threw a beam of red light in a spike upon its shining by ordinary light emanating from the flashlamp. A detailed scientific account of Maiman's experiment is provided in Chap. 4. It is not surprising that the birth of lasers became headline news across the world with captions such as "L.A. Mans's Light Ray Outshines Sun" or "Light Brighter Than Sun Can Plant 10-Mile-Wide Beam on Moon." Controversies, however, exist even today with regard to the allocation of credit on inventing lasers to a score of researchers, primarily American, who competed fiercely in the race to build the first laser [6, 7]. Although Maiman is largely credited with the invention of the first laser [8], Charles H. Townes (1915–2015) shared the 1964 Nobel Prize in Physics with two Russian physicists Alexander Prokhorov (1916–2002) and Nikolay Basov (1922–2001) for the development of laser theory and construction of a Fabry-Perot-cavity [9] based light oscillator, considered the backbone of the laser. One-half of the 1981 Nobel Prize was awarded to two other contenders Nicolaas Bloembergen (1920–2017) and Arthur Leonard Schawlow (1921–1999), as a consolation, albeit for the development of laser spectroscopy. Another strong contender Gordon Gould (1920–2005), who incidentally coined the word laser, won a long battle in an American court in 1973, invalidating a patent earlier issued jointly to Townes and Schawlow and letting him secure a series of patents on laser concepts and applications. It was at least a financial solace to Gould to eventually earn millions of dollars in royalties through these patents, a sum far more than what the Nobel Prize would fetch. Obviously dismayed for being ignored by the Nobel committee, Maiman took the recourse to write *The Laser Odyssey* [10] largely a biography of his illustrious scientific career, and spared no effort venting his annoyance in missing out on the coveted Nobel Prize.

The first laser flashing its light fired the imagination of scientists across the world. Unfortunately, Einstein did not live to witness the invention of laser, the seed of which he planted way back in 1917. That at the time of its birth, a laser was suggested to be brighter than the Sun was not a shot in the dark. Even a laser pointer with barely a milliwatt of power can be much brighter than the Sun. The highly directional emission from the laser, an outcome of the integration of the cavity with stimulated emission, has made this possible. As a matter of fact, it is the brilliance of the cavity that has endowed a laser with directionality and monochromaticity: the two key features that distinguish it from a conventional light source. Dubbed in the beginning as a solution in search of problems, today the laser has pervaded almost every walk of our life. Its otherworldly applications have grown rapidly over the years and are still growing. It has left an indelible mark and continues to do so in areas of direct relevance to human life, e.g., health, security, science, education, industry, transport, entertainment, communication, energy, and almost anything that one can think of. The fact that numerous Nobel Prizes have been awarded to brilliant pieces of work that involve laser in one way or another also bears testimony as much to its Midas touch on the research areas it has come in contact with as its role in the advancement of science.

Today, we have a wide variety of lasers giving out light right through the far-infrared to vacuum ultraviolet region of the electromagnetic spectrum. The size and power of the lasers can also vary widely. The semiconductor laser, also called diode lasers, which is key to optical telecommunication, optical data storage, medical applications, and many more, can be tinier than even the breadth of a human hair [11]. Fabrication of such a tiny device that is capable of giving out directed coherent light even at room temperature has changed our world forever. Herbert Kroemer (b–1928) and Zhores Alferov (1930–2019), the inventors of this incredible laser, were awarded the Nobel Prize in Physics in 2000. Miniaturization of semiconductor lasers is key to the development of integrated photonics, the building block of next-generation optical communication devices.

It is also true that many tiny diode lasers, when put together, can power a fiber laser, the active length of which can run into many kilometers, even though they are usually coiled into a convenient size. The ability of these ultralong lasers to yield high-quality optical beams with continuous power running into several to tens of kW has led to their emergence as an industrial workhorse worldwide. When such a diode pumped fiber performs as an optical amplifier, it has the ability to directly amplify a feeble optical signal at the intermediate booster stages in long-distance national, international, and transoceanic fiber-optic communication networks. All-light fiber-optic communication, bypassing the erstwhile requirement of intermediate conversion of light into electronic form and back to light again, has now revolutionized telecommunication and Internet connectivity.

A laser can also be as large as the combined size of three soccer fields [12], comprising 192 lasers and approximately 40,000 optics for their operation and combining their emissions into a single colossal pulse of light lasting only for an extremely tiny fraction of a second. Scientists believe that this huge laser facility, popular as the National Ignition facility (NIF) built at the Lawrence Livermore National Laboratory, California, may be capable of creating the condition that prevails at the core of our Sun. To say that in a flitting billionth of a second this astounding laser system delivers coherent light of energy dwarfing the average power that the whole US power grid is capable of generating is not a leap in the dark. This gigantic effort is primarily aimed at harnessing the power of fusion to manufacture clean energy for humankind in the same way as the Sun does for the entire solar system[7]. In contrast, many of us may be familiar with the handy laser pointer that sits so well into our hands. Then, there are those other kinds of lasers that can fit in a box of the size of a refrigerator and can slice through hard metal as if it were cheese.

From the modest beginning of light pulses of millisecond duration emitted by a ruby laser, we have now come a long way; there are lasers that are capable of twinkling light for barely a million billionth of a second, i.e., 10^{-15} seconds or a femtosecond. A deeper understanding of laser dynamics and cavity physics in conjunction with stimulated emission has led to the conceptualization of techniques such as Q-switching, cavity dumping, mode locking, or colloidal pulse mode locking. The implementation of these techniques in the operation of a laser has allowed pulse compression to gradually progress through micro, nano, and up to the point of femtosecond and even beyond. A flash of light lasting for about a femtosecond may appear to be ridiculously inconsequential, but it can, as a matter of fact, accomplish seemingly unthinkable tasks. It can function as a scalpel of extraordinary sharpness,[8] offering a surgeon the luxury of performing a high-precision job involving delicate organs such as the eye and heart or allowing a photochemist to take a snapshot of the intermediate species formed during a chemical reaction defeating the lightning speed of their formation. A femtosecond pulse is a communication engineer's dream, as it presents seemingly limitless bandwidth[9] for data stuffing. Furthermore, to possess terawatt average power, the energy content of a femtosecond

---

[7] On December 5, 2022, the researchers at the NIF achieved what eluded them for decades- 'Self-sustaining fusion that can yield more energy than goes into it'. In a nutshell, this feat, regarded as the most remarkable scientific achievement of the 21st century, forms a crucial milestone in humans' quest to replicate the process that powers the Sun and the stars of the universe.

[8] As a femtosecond laser pulse is focused on a material surface or a human tissue, the electrons acquire energy from the incident radiation, but the time is too short for them to transfer it to the lattice. The electrons, in turn, become extremely hot and get out of the surface taking along a chunk of material through Coulomb attraction. As there is no local heating, there is no heat affected zone, and consequently, such a process is termed as cold ablation. Such ultrashort pulses are therefore ideal for microsurgery and micromachining applications.

[9] This is a direct consequence of Heisenberg's uncertainty principle that can be mathematically expressed as $\Delta \nu \times \tau \approx 1$, where τ is the duration of the pulse and $\Delta \nu$ is its bandwidth. Thus, the shorter the pulse, the broader the bandwidth.

pulse needs to be only 1000th of a Joule in complete contrast with the kilojoule level of energy requirement at NIF. Although a terawatt femtosecond pulse may not ignite a fusion reaction, it, when appropriately focused, can recreate the extreme condition that prevails at the center of a star, a researcher's dream, right on the table top in the laboratory. However, the power of the ultrashort pulse straight out of the laser is barely a kilowatt and can hardly be boosted by letting it pass through a light amplifier, as its growing power will trigger irreversible optical damage to the amplifier itself. This seemingly insurmountable problem was circumvented by Gerard Mourou (b–1944) and his doctoral student Donna Strickland (b–1959) by cleverly employing the chirped pulse amplification (CPA) technique. They first stretched the femtosecond pulse by many folds to bring down its power, allowing its amplification without the risk of destroying any amplifying material, and then compressed it back to its original temporal state. This made possible for a kilowatt ultrashort light pulse to coolly leap to the level of terawatt or even beyond. Creation of an ultrashort laser with the ability to deliver power of such colossal magnitude is often regarded as the second birth of the lasers. The CPA technique basically added a new dimension to the laser technology and led to the emergence of an area of research known in the common parlance as "Extreme Science with Extreme Light." The 1985 inventors of this technique were eventually honored with the coveted Nobel Prize in Physics in 2018.

The infinitesimal momentum carried by the photons constituting a laser beam can be exploited through the reflection or refraction of light to manufacture a force so tiny as to be able to trap or maneuver the motion of micro-objects such as viruses, bacteria, living cells, atoms, molecules, and even strands of DNA in a tweezer-like manner. This remarkable ability of lasers has had such a profound impact, particularly in the areas of biology, medicine, nanoengineering, and nanochemistry, that the originator of this novel concept, Arthur Ashkin (1922–2020), was awarded the Nobel Prize in Physics in 2018. He was 96 years old at the time of the award and became the oldest person to receive a Nobel Prize.[10]

Just as in the manufacturing of laser fingers to grab tiny particles, the tiny momentum of a light photon can also be exploited to cool an atom that moves about at breakneck speeds at room temperature. By shining the atom with multiple laser beams from all sides, its momentum can be bumped at the expense of the photon's momentum. As an atom absorbs a photon traveling in a particular direction and reemits it in a random direction, its momentum reduces. By making the same atom absorb multiple times, it can be cooled to extremely low temperature and captured in a trap. Laser cooling and trapping have applications in diverse areas ranging from atomic clocks, atomic interferometers, and spectroscopy to quantum physics involving, in particular, the Bose-Einstein condensate. Steven Chu (b–1948), Claude Cohen-Tannoudji (b–1933), and William Daniel Phillips

---

[10] The distinction of being the oldest person to be awarded the Nobel Prize was, however, extremely short-lived for Ashkin as in the following year (2019) Nobel Prize in Chemistry was awarded to 97-year-old John Bannister Goodenough (b–1922).

(b–1948) were awarded the Nobel Prize in Physics for independently conceptualizing and developing the laser cooling technique.

The deadly heat rays that H. G. Wells (1866–1946) fictionalized Martian invaders to have possessed in his novel *The War of the Worlds*, written in 1896 long before even the conceptualization of stimulated emission, let alone laser, are no longer fantasy. Today, there are lasers powerful enough to zap an incoming enemy missile or destroy an airplane in flight. The use of lasers in the military, both as a weapon and guide, is well known and is being actively pursued by many advanced nations in the world, the details of which, being classified, are generally not available in the open literature. It is worth mentioning in this context of an ambitious plan, nicknamed the Star Wars program, that America undertook in the Cold War era during the presidency of Ronald Reagan. The basic strategy was to shield its territory by placing high-power lasers in strategically located orbits with the capability of striking and destroying a nuclear warhead in the flight. The idea was basically to destroy the missile in the boost phase on the sky of the country of its origin. The cessation of the Cold War in the early 1990s, however, brought a premature end to this program about a decade after its inception.

Lasers have had a profound impact on the exploration and harnessing of space. As you read this, there may be one or another spaceship equipped with advanced laser communication devices, surging forward even further into our solar system. Notwithstanding the vast distances involved, the pinpoint precision of laser-based communication will help gather science much faster as the vehicle lands on the surface of another planet. A massive effort is now underway to place several thousand satellites into low-altitude orbits to form a constellation. All the satellites are to be linked via laser, and when fully operational, the constellation is expected to provide Internet connectivity across the globe, including rural and geographically isolated areas.

The laser has its footprint everywhere from the infinitesimal to the astronomical – the largest conceivable reality – the expansion of the universe. An informed reader will be aware of the iconic LIGO[11] (Laser Interferometer Gravitational-wave Observatory) experiments designed to unlock the most secret door of nature, the unification of all its forces, something that eluded Einstein even up to the point of his falling asleep that final time.[12]

---

[11] LIGO is a large-scale physics observatory located in the USA with primary aim of detecting gravitational waves by employing the technique of laser interferometry. It is of interest to note here that 2017 Nobel Prize in Physics was awarded to Rainer Weiss (b–1932), Barry C. Barish (b–1936), and Kip S. Thorne (b–1940) for the very first LIGO-based detection in 2015 of the Universe's gravitational wave that originated from a collision between two black holes.

[12] Einstein died in the early hours of April 18, 1955, and on the previous night before he went to sleep that last time, he persevered with his obsession, a unified field theory that would tie together all the forces of nature. Twelve pages filled with equations found on the table adjacent to his hospital bed bore testimony to this.

The above is only a glimpse from the world of incredible lasers and some of the mind-boggling effects that they exhibit, and this book is intended to reflect the epic journey that lasers have undergone to date. The three major aspects that would undoubtedly showcase a laser most comprehensively are the principle of working of a laser, the role it plays in the advancement of science, and the influence it wields in our life. The book has been planned to unmask all these three sides of the lasers by employing an engaging style that combines lucid text with explanatory illustrations and anecdotes picked up from history. Our intention of adopting an intuitive and largely nonmathematical approach, instead of the typical pedagogical style, is primarily to make the text readily palatable to undergraduate students as well as catch the attention of a wide cross section of readers both within and outside the laser community. Quite aptly, therefore, the book has been named *A Beginner's Guide to Lasers and Applications* to be published in two parts: Part-1, *Insights into Laser Science*, and Part-2, *The Laser's Impact on Science and Humanity*.

A Brief Description of the Contents of Part-1: Spread into 12 chapters, inclusive of the introduction, that forms Chap. 1, it broadly dwells on what a laser is, how it works, and the uniqueness of the light it gives out. Chapters 2 and 3 simply recall the basic properties of light and the basic concepts of discretization of energy levels, respectively, the two major precursors to understanding lasers. There is no denying of the fact that light and its actions are profoundly interlaced with our day-to-day life. Consequently, real-life examples, terrestrial or celestial, are often chosen to allow fathoming incredible light with ease. Likewise, to strike a sense of energy quantization of microscopic particles to the reader's mind, its manifestation in the macroscopic world, we are familiar with, is brought to the fore. A largely standalone chapter on lasers that follows next has been planned as a prelude to an in-depth understanding of laser fundamentals. Entitled "Lasers: At a Glance," this chapter is intended to impart rudimentary knowledge to the readers to prepare them to fathom a laser more intelligibly as they read through the latter chapters of this volume dealing with the physics of laser more intricately. In addition to providing deeper insight into the concept of population inversion, Chap. 5 also enlightens readers on the inherent difficulties in its realization and some practical ways to overcome them. A significant part of the chapter is also devoted to identifying a medium most suitable for effecting and sustaining the condition of population inversion and, in turn, lasing. Chapter 6 begins with a series of simple but revealing illustrations to justify why there would not be a laser without a cavity. A few thought experiments, cleverly planned with equally thoughtful illustrations, have been used to decode the rich physics that a cavity holds. An easy-to-understand explanation of the deeper concepts involving a cavity such as its ability to transmit resonant light, the stability criteria, and its operation in Gaussian, multi-transverse, or unstable modes is another highlight of this chapter. A laser can be operated both in the continuous and pulsed modes, and as a matter of fact, these two operations complement each other quite remarkably as far as their applications are concerned. Chapter 7 has been planned to impart a comprehensive knowledge on the working of both continuous and pulsed lasers. In addition to providing a deeper understanding of the two kinds of gain broadening, homogeneous and inhomogeneous, Chap. 8 has also been planned to reveal the strong bearing it has on the operation of a laser. Occurrences of spatial and spectral holes in the gain and their manifestations in both standing and travelling wave cavities are some of the major highlights of this chapter. Elucidating the gainful exploitation of the indelible trails it leaves on the laser dynamics, such as the Lamb dip in the operation of a single mode gas laser, has also been a major endeavor of this chapter. Pulsed operation of a laser is central to a wide variety of applications, and understandably, a great deal of research has been devoted to boosting the performance of a pulsed laser. Chapter 9, in addition to providing deeper physical insight into the various techniques to achieve pulse compression, also describes the challenges of amplifying an ultrashort pulse and chirped pulse amplification as a remedy. Chapter 10 comprises a survey of a few specific laser systems. For this purpose, the lasers have been grouped in terms of the state of the active medium, viz., gas, liquid, and solid-state lasers, and the operation of a popular laser from each category has been described with a special emphasis on the underlying physics. The active medium of a gas laser can be atoms, ions, and molecules. Consequently, we choose *He-Ne*, argon ion, and CO_2 lasers from these respective categories for this survey. While dye lasers represent the liquid laser, Nd-YAG and Ti-sapphire lasers have been chosen, respectively, as the discrete wavelength and continuously tunable lasers from the solid-state laser category. The other kinds of lasers, the working of which are distinctly different, have been addressed separately in this chapter. They include free electron lasers, excimer lasers, chemical lasers, gas dynamic lasers, and fiber lasers. As basic knowledge of molecular spectroscopy and semiconductor physics is a prerequisite to understand the working of CO_2 and semiconductor lasers, respectively, their operations have been described in two different chapters. The constituent atoms of a molecule can vibrate and also rotate at the same time. Similar to the electronic excitation energy, the molecular, vibrational, and rotational energies are also quantized. The basics of molecular spectroscopy, introduced at the beginning of Chap. 11, will allow comprehending the intricacies involving the operation of a CO_2 laser, one of the most well-researched, powerful, and versatile lasers. The underlying physics of semiconductors, beginning from the conceptualization of holes to the emission of light from a p-n junction, has been brought out in an engaging style in the beginning of Chap. 12. The latter half of this chapter provides illuminating insight into the working of semiconductor lasers with a particular emphasis on heterojunction, both edge and surface emitting types, quantum well, and quantum cascade lasers.

Whenever possible, the background information has been provided as a footnote lest it disrupts the flow of the text. Preference of footnotes over referencing kept the number of references to each chapter to a bare minimum. This, we believe, will make the reading more enjoyable and seamless.

Chapter 2
Behavior of Light

2.1   Introduction

To state that "light was the key for the evolution of life on our planet" is anything but
an overstatement. The process of photosynthesis is known to have been established
more than two billion years ago, once the temperature of Earth became cold enough
to form neutral atoms allowing sunlight to penetrate deep into its atmosphere. Prior
to this, the temperature was too high, and the matter was predominantly in the state
of plasma through which light could travel only a short distance.[1] This initially
inhibited the process of photosynthesis. The large quantity of oxygen being
manufactured in the atmosphere as a by-product of photosynthesis not only facilitated evolution of oxygen-breathing life but also kept the atmosphere breathable ever
since. Light played a pivotal role not only in the creation of life but also in the
evolution of human thoughts over centuries that shaped our understanding of nature.
We have come a long way today from the humble beginning when Sun was our only
source of light. From the modest camp fires used by our cave-dwelling ancestors, we
gradually started to use oil lanterns. The discovery of electricity brought about a
revolution in the development of artificial sources of light. The rudimentary incan-
descent light has now progressed through fluorescent light to the more sophisticated
LED (light emitting diode) light. Finally, the conceptualization of the stimulated
emission of light eventually led to the invention of laser light, an amazing gift of
science to humankind. By mastering the techniques of controlling and transporting
light, we, from the overly modest beginning of using bonfire to cook food, are now

---

[1] Plasma is a gas of ions and electrons and possesses a natural frequency of oscillation that increases
with the density of the charged particles. It is a common knowledge that a body of plasma allows
electromagnetic waves of frequency above its own frequency to pass through and blocks the rest.
The density of plasma, in the early Universe of extreme temperature, was much too high and so was
also its natural frequency. Sunlight of even high frequency, let alone the lower ones, could,
therefore, barely leak through the plasma during this period.

© The Author(s), under exclusive license to Springer Nature Switzerland AG 2023
D. J. Biswas, *A Beginner's Guide to Lasers and Their Applications, Part 1*,
Undergraduate Lecture Notes in Physics,
https://doi.org/10.1007/978-3-031-24330-1_2

able to employ the power of light in a much grander way. Needless to say, harnessing the seemingly endless potential of light, which has intrigued the human mind from time immemorial, has been crucial to the advancement of civilization. It is well known that the development of light-based instruments has contributed over the years toward enriching our knowledge. That the observation of planetary motion through a telescope made it possible for Galileo to perceive the heliocentric universe, annulling irrefutably the superstition of geocentric notions that hindered the flow of knowledge era after era, is now history. To say that laser is second to none in harnessing the power of light will be an understatement, as it is miles ahead of any other to say the least. Elemental knowledge of the nature and properties of light is an essential prerequisite for understanding a laser. The content of this chapter has been specifically planned to not only impart a basic knowledge of light and its behavioral pattern and properties but also to kindle an enthusiasm in the mind of the readers. Readers with a background in optics may also consider reading through this chapter to refresh their knowledge on light and optics. As a prelude, a brief historical review has been provided on the voluminous efforts expended over centuries toward understanding light.

2.2 Fathoming Light: A Historical Glimpse

The seemingly mystical nature of light has fired the imagination of some of the finest scientific minds dating back from the primeval era leading to the gradual unraveling of the complexity enveloping it. The misconception that light rays emitted by our eyes allow us to perceive an object, advocated by Plato (BC 428–328), one of the chief architects of the ancient Greek civilization, and his followers, persisted for over 1000 years. Alhazen (965–1040), who belonged to the Islamic golden era[2] and is often regarded in the West as the greatest physicist between Archimedes and Newton, was the first to conclusively prove through credible scientific experiments that light rays originate from a source and not from the eye. He completed writing a book on optics in 1027 providing a comprehensive treatment of the theory of vision, perception of color, the lens, and the magnification of an object originating from bending of light due to refraction on a curved surface. A Latin translation of this book became available sometime around the end of the twelfth century and would remain a well-referred document on optics until Newton published *Opticks* in 1704. René Descartes (1590–1650), considered a key figure in the Scientific Revolution,[3] documented several topics relating to the properties of light including the laws of

---

[2] Islamic golden era is known to have spanned between eighth and fourteenth centuries when Islamic civilization greatly flourished culturally, scientifically, and economically.

[3] Scientific revolution that primarily took place in Europe during seventeenth and eighteenth centuries led to a proliferation of scientific thoughts transforming our understanding and views on nature.

Fig. 2.1 The cover page of
the third edition of Opticks
that was published in 1721
and contained Newton's
name. The first and second
editions didn't have his
name on the cover

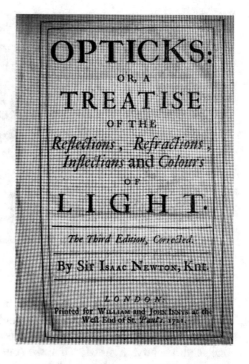

reflection and refraction in a book of optics, *Dioptrics*, published in 1637. Dutch
scientist Willebrord Snellius (1580–1626) is, however, credited with formulating the
laws of refraction that are well known in the English speaking world as *Snell's Law
of Refraction*. It is of interest to note here that Pierre de Fermat (1601–1675), a
French mathematician, was the first to introduce the principle of least time in optics,
a manifestation of the rectilinear propagation of light, and used the same to offer a
correct derivation of Snell's law. In 1690, Christian Huygens (1629–1695), a Dutch
polymath,[4] introduced the wave picture of light for the first time, which would
emerge as a competitor of Newton's corpuscular theory, in his book on *Treatise on
Light (Traite de la Lumiere)*. It is at this juncture that Sir Isaac Newton (1642–1727),
a central figure in the history of science, made his appearance in 1704 through his
book *Opticks: or, A Treatise of the Reflexions, Refractions, Inflexions and Colors of
Light*. The book is chiefly a record of the results from ingeniously planned experiments and made a profound impact on the field of optics (Fig. 2.1). He conclusively
established that sunlight is colorless or white and is made up of seven different
colors, viz., red, orange, yellow, green, blue, indigo, and violet, and was able to
decompose these colors by letting sunlight travel through a prism (Fig. 2.2). He is

---

[4] A polymath is an individual whose expertise spans over a significant number of disciplines.

Fig. 2.2 A beam of white light undergoes reflection and refraction as it enters a glass prism. The prism disperses the white light into its constituent colors. (Credit – Spigget, Wikimedia Commons)

credited with the development of the particle or corpuscular theory[5] of light and was able to interpret most of the phenomena exhibited by light such as reflection, refraction, polarization, and dispersion by invoking this. Newton's authority in the scientific world was so pronounced that nobody could challenge the corpuscular theory, and Huygens wave theory of light took a back seat until the beginning of the nineteenth century. In 1801, Thomas Young, a British physicist, performed his famed double slit experiment, considered one of the all-time ten most beautiful experiments in physics [13], revealing the occurrence of interference of light as a direct manifestation of its wave nature. Through this experiment, Young reestablished the wave theory of light, originally proposed by Huygens, overcoming the century old belief of its corpuscular nature. It is remarkable that exactly another century later, the particle nature of light made a complete turnaround when in 1905, a year of scientific miracles, Einstein (1879–1955) was able to resolve the puzzle of the photoelectric effect by considering light to be composed of a stream of photons. Taking a cue from the fact that light can manifest itself both as a particle and as a wave, Louis de Broglie (1892–1987), a French physicist, proposed in 1924 that matter can also behave both as wave and particle. He was awarded the Nobel Prize in Physics in 1929 for this groundbreaking work that planted a seed of quantum mechanics to describe the physical properties of atomic and subatomic particles constituting the microscopic world. It is also quite remarkable that another interference experiment, but now with an electron,[6] arguably became the all-time number

---

[5] Corpuscular theory of light, proposed by Newton in the early eighteenth century, states that light is composed of tiny particles, called corpuscles, which propagate in straight line with finite velocity and hence possess momentum.

[6] It is of interest to note here the thought double slit experiment with a single electron that Nobel Laureate Richard Feynman described in now-famous series of lectures at the California Institute of Technology in 1965. Furthermore, the advances in the manufacturing techniques made possible the real-world demonstration of this famed thought experiment several decades later.

Fig. 2.3 A captivating image of the sky painted magically with colors spilling out of the fading light of the setting Sun off the west coast of Portugal. (Credit -Alves Gasper, Wikimedia commons)

one physics experiment [13]. An informed reader will be aware that the outcome of this experiment, revealing the wave nature of electrons, placed quantum mechanics on a firm footing – but that's another story.

2.3   Rectilinear Propagation of Light

The knowledge that light follows a straight path and does not bend on common objects, we routinely encounter, is indeed primordial. Our cave-dwelling ancestors knew to block light with opaque barriers to preserve their privacy. It is a common knowledge that while we can hear a person in another room, we cannot see him or her. A typical household object, such as the frame of a door or window, can bend sound but not light. If light, instead, would have followed a curvilinear path, there would not be occurrence of day and night or lunar and solar eclipses, or formation of shadows, for that matter. The straight-line motion of light has manifested in a variety of phenomena such as reflection, refraction, dispersion, and scattering. The display of a rich array of colors in the sky during sunrise or sunset (Fig. 2.3) and the exquisite blue sky or the magnificent rainbow (Fig. 2.4) that have mesmerized the human mind from time immemorial are all a direct consequence of these. To gain insight into these phenomena, we first need to conceptualize a ray of light. We now know that a luminous source gives out light that spreads all around it in straight paths. A point

Fig. 2.4 The Rara Lake, a famed tourist spot of Nepal, appears even more spectacular when a rainbow, formed on the horizon, embellishes the magnificent sky at the background (Credit – Chandra Chakradhar, Wikimedia Commons)

Fig. 2.5 Propagation of light from a point source depicting what is called a beam and a ray. Source emits light all around it, but only an array of rays shining onto the barrier is shown here

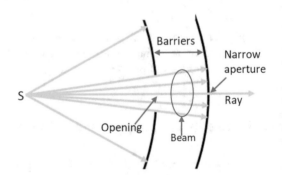

source can, thus, be thought of sending out rapidly expanding light of spheres. It is adequate to consider the passage of a part of the sphere through an aperture described in Fig. 2.5. What emerges through the aperture is called a beam of light. It is common knowledge that with the help of one or more optical elements, the beam of light can be made converging, diverging, or parallel. It is often convenient to understand the behavior of light by considering a beam so narrow that it can be thought of as a straight line of zero width termed a ray of light (Fig. 2.5). Understandably, a ray of light possessing only length, but no width, can neither converge nor diverge and forms the basis of understanding each and every phenomenon that light exhibits.

2.3.1   Reflection of Light

Whenever light is incident on any surface, a part of it, small or large, always gets reflected back. In this context, we need to consider three different kinds of surfaces, viz., (1) a smooth or polished surface such as that of a plane mirror; (2) an opaque and rough surface, abundantly present in nature and normally characterized by micro-irregularities, such as the page of this book you are reading through; and (3) the surface (interface) separating two optically transparent media such as air and water or air and glass. The reflection of light occurring from all these three different kinds of surfaces is illustrated in Figs. 2.6, 2.7, and 2.8 along with their real-world manifestations. In the case of reflection of a parallel beam of light off a mirrorlike surface, the incident rays (such as AO) and the reflected rays (such as OB) make exactly the same angle with ON, a normal to the reflecting surface (Fig. 2.6). A reflection of this type, where the angle of incidence equals the angle of reflection (i.e., $\angle AON = \angle BON$), is called specular reflection. Furthermore, the reflected ray

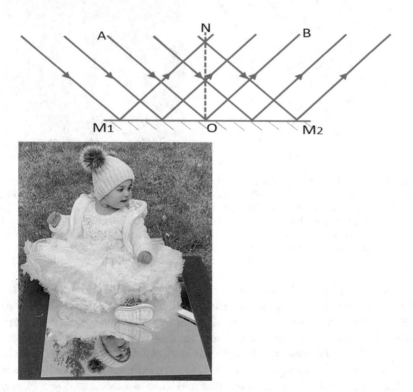

Fig. 2.6 Specular reflection of light, which obeys the laws of reflection, occurs on a polished surface like that of a mirror. A bright and sharp image is often formed from such a reflection. An example is shown here for an object, the expanse of which is not confined to the smiling toddler sitting on the reflector but stretches as far as the spotless blue sky. (Credit: Debosmita Biswas, University of Washington, Seattle)

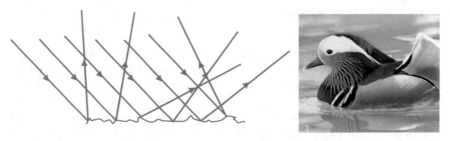

Fig. 2.7 A magnified view of an irregular surface and a parallel beam of light shining upon it. Reflected rays travel in all possible directions as the corresponding incident rays strike the micro-irregular surface at widely varying angles. This diffuse nature of reflection makes it possible for us to perceive the endless beauty of nature. A case in point is this exquisite Mandarin duck shown alongside. Day light shines onto this magnificent creation and is reflected from every conceivable point of its body part revealing, in turn, its dazzling colors to our eyes from any position

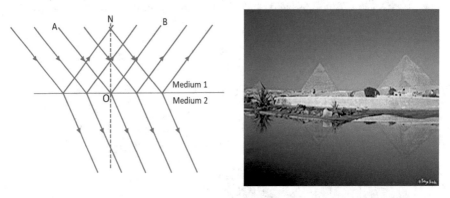

Fig. 2.8 Schematic illustration of specular reflection at the interface between two media when light travels from one (medium 1) to the other (medium 2). A part of the incident light is reflected back into "medium 1," while the remaining light travels into "medium 2." Reflection of light from air-water interface is very common in nature. An example, the sunlit Giza Pyramid, the oldest of the seven wonders of the ancient world, and its image formed from reflection off the still water of Nile River, is shown in the adjacent picture (Credit: Turizm-Art, Wikimedia Commons)

(OB) always lies in the plane defined by the incident ray (OA) and the normal (ON). The image formed as a result of specular reflection of a typical object is placed next to it. In contrast, when a parallel beam of light shines on an irregular surface, as is the case with most natural objects, different incident rays will meet the surface at different angles and, thus, will be reflected in all possible directions. A magnified view of such a surface and reflection off it is shown in Fig. 2.7. Such reflections are, therefore, diffuse or non-specular in nature and allow us to view an object from any position revealing, in turn, its uniqueness. When light travels from one medium to another, reflection also occurs at the interface when a part of the light returns to the incident medium. The rest of the incident light bends while entering the second medium. Specular reflection can occur in the case of an absolutely flat and blemishless interface and is schematically illustrated in Fig. 2.8. The formation of images on the air-water interface is a very common sight in nature and owes its origin to the reflection of this kind. The adjacent picture here captures an irrefutably sharp image of the Great Pyramid of Giza formed as a result of the reflection of light from an absolutely calm and wriggle-free air-water interface of the Nile River. The image is also unmistakably less bright compared to the sunlit pyramid and bears the signature of the fact that only a part of the light from the object, which is incident on the water surface, is reflected back into the air to reach our eyes. In contrast, almost the entire amount of incident light is reflected from the surface of a mirror or polished metal, and the image can, therefore, be just as bright as the object, as evident from Fig. 2.6. The mirror does not allow any light to escape, and consequently the image, be it that of my granddaughter, Coco, sitting on the mirror, or the tree branch hanging atop or the sky for that matter, is just as bright as the object itself. You may, as a matter of fact, be wondering if the blue hue of the sky behind the mirror is a shade deeper than the real sky hiding in the front!

French physicist Augustin-Jean Fresnel (1788–1827) was the first to provide physical insight into the reflection of light from the interface between two media, known as Fresnel's reflection. He analytically established an expression for the fraction of light (R) reflected from such an interface revealing its critical dependence on both the angle of incidence and the refractive index[7] of the two media. In the case of a normal angle of incidence, R can be expressed as

$$R = \left[\frac{\mu_1 - \mu_2}{\mu_1 + \mu_2} \right]^2 \tag{2.1}$$

where μ_1 and μ_2 are the refractive indices of medium 1 and medium 2, respectively. For an air-glass interface for which $\mu_1 = 1$ and $\mu_2 = 1.5$, R assumes a value of approximately 4% at the normal angle of incidence. The Fresnel reflection has a strong bearing on the construction of lasers. Operation of a laser, as we would see later, requires enclosing the lasing medium between two parallel mirrors of high reflectivity. Glass and many of its variants, being transparent across a wide wavelength range covering near-infrared through visible to ultraviolet, have emerged as the preferred material for making such mirrors. The small Fresnel reflectivity at the air-glass interface, however, poses a major challenge. We shall soon see, as we read through this chapter, how thin film technology can exploit the phenomenon of interference of light to manipulate the value of the Fresnel reflectivity to the required level.

---

[7]Light travels slowly in glass or water compared to air or vacuum. The electrons present in an optically transparent medium interact with the electric field ubiquitous with light and, in turn, slows it down. The refractive index (r.i.) is an optical property of a medium that basically relates to its power of slowing down light and is expressed as the ratio of the speed of light in vacuum to that in the medium. A medium with higher r.i. is called optically denser with respect to the other, while a medium with lower r.i. is called optically rarer. Light travels faster in water than in glass, so water is optically rarer than glass and conversely glass is optically denser than water.

Fig. 2.9 A retroreflector is
a combination of two
mirrors inclined to each
other at right angle.
Reflected light here always
travels in a direction exactly
opposite to the incident
light. This property has been
underlined here for two rays
of light striking the reflector
at different angles

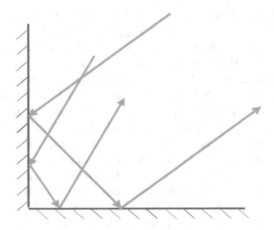

2.3.1.1 Retroreflection of Light

The structure of some surfaces may be such that the reflected light travels exactly in a
direction opposite to that of the incident light. The reflection exhibited by such a
surface is called retroreflection. A simple retroreflector can be constructed by placing
two ordinary mirrors mutually perpendicular to each other (Fig. 2.9). It is straightforward to show by way of using laws of reflection and simple geometry that light
incident from any direction on such a reflector will always travel back exactly in the
opposite direction. We shall see in a latter chapter (V-II of this book) how the
positioning of a retroreflector on the surface of the Moon[8] made lunar ranging by a
laser possible to an incredible accuracy. It may be of interest to note here that certain
animals' retinas act as a retroreflector to basically help them navigate in the night.

2.3.2 Refraction of Light

When light is incident on an interface between two transmitting media, a part of the
light is reflected back into the incident medium, and the remaining light exhibits
bending before traveling into the second medium. The phenomenon of bending of
light, as it travels from one transmitting medium into another, at the interface is
called refraction of light and is schematically represented in Fig. 2.10. Willebrord

---

[8]David Scott, an American astronaut, who commanded the Apollo 15 lunar mission had placed a
retroreflector on the surface of the Moon on July 31, 1971. The unique property of such a reflector to
return the reflected beam in the same direction as the incident beam subsequently allowed the
measurement of distance between the Earth and Moon to an incredible accuracy of 2 cm, i.e.,
equivalent to an error less than even 1 part in 20 billion. Incidentally Scott is one of four surviving
Moon walkers today.

Fig. 2.10 A part of the
light, as it shines on an
interface between two
transparent media, is
reflected back into the first
medium, and the remaining
part is refracted into the
second medium after
experiencing a bending at
the interface. In case of a flat
and smooth interface, the
incident, reflected, and
refracted rays and the
normal NON' lie on the
same plane

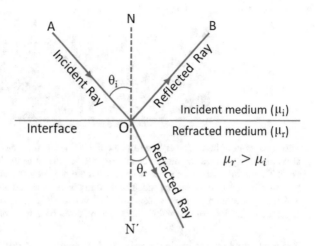

Snellius was the first to connect the bending of light at the interface with the optical
properties of the media through the following equation, known as Snell's law:

$$\mu_i \sin \theta_i = \mu_r \sin \theta_r \qquad (2.2)$$

Here μ_i and μ_r are the refractive indices of the incident and transmitting media,
respectively, and θ_i and θ_r are the angles of incidence and refraction, respectively. As
the value of $sin\ \theta$ increases with θ in the first quadrant, it readily follows from Snell's
law that light bends toward the normal while travelling from a rarer to denser
medium. On the other hand, light, while travelling from a denser to rarer medium,
will bend away from the normal. This gives rise to the prospect of the entirety of the
light returning back to the incident medium from the interface without exhibiting any
refraction, a phenomenon called the total internal reflection of light.

2.3.2.1   Total Internal Reflection of Light

To gain insight into this seemingly interesting behavior of light, we analyze the case
of light traveling from a denser medium such as water to a rarer medium such as air
as shown in Fig. 2.11. The trajectories of four light rays emanating from a point P in
the water and striking the interface at progressively increasing angles of incidence
are illustrated in this figure. The angle of incidence of the first ray is relatively small,
and it understandably bends away from the normal after refraction. The second ray
with a relatively higher angle of incidence bends further away from the normal
following refraction. The angle of incidence of the third ray is such that it simply
grazes over the interface after refraction. The angle of incidence for which this

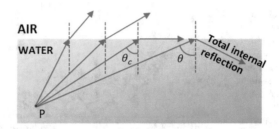

Fig. 2.11 Four light rays emerging from a point "P" inside the water strikes the water-air interface at progressively increasing angles of incidence. The normal to the interface are shown by dotted lines. The first ray expectedly bends away from the normal after refraction, while the second ray bends further away from the normal. The third ray is incident at an angle, called critical angle, for which the refracted ray simply grazes over the interface. The fourth ray striking the interface at an angle more than the critical angle is internally reflected back into water

happens is called the critical angle[9] and is usually denoted by θ_c. Understandably, if the angle of incidence exceeds θ_c, light has to return back to the incident medium as is the case with our fourth ray. Such a phenomenon that can occur only when light travels from an optically denser to a rarer medium is called the total internal reflection of light. We will find in a latter chapter (V-II of this book) as to how the exploitation of total internal reflection in conjunction with lasers has caused a revolution in the area of optical communication and, in turn, changed our world forever. Exploitation of reflection and refraction experienced by a beam of laser can lead to the generation of an extremely tiny force capable of maneuvering the motion of micro-particulates. As we will find in another chapter of next part, this remarkable phenomenon has opened up an altogether new branch of optics, called optical tweezers.

2.3.2.2 Dispersion of Light

Rainbow, the multicolored exquisite arc of light thrown on the sky when the Sun shines on raindrops, has, over the years, fired the imagination of the thinkers, the scientists and poets alike. Aristotle (BC 384–322), the Greek philosopher, was the first to perceive in BC 350 a connection between the rainbow and the water droplets present in the atmosphere [14]. No major advances occurred until 1637 when René Descartes (1596–1650), a French born Dutch scientist, suggested that the multiple color of the rainbow owes its origin to the splitting of sunlight by the raindrops. The enigma of rainbow was finally cracked in 1672 when Newton proposed the theory of dispersion of light. He showed that sunlight is white and is made up of seven colors,

---

[9] At the critical angle of incidence, θ_c, the angle of refraction is $90°$. Snell's law of refraction therefore yields

$$\mu_i \sin \theta_c = \mu_r \sin 90^o \Rightarrow \theta_c = \sin^{-1} \frac{\mu_r}{\mu_i}.$$

Fig. 2.12 Schematic illustration of spatial splitting of white sunlight upon refraction on an air-glass interface

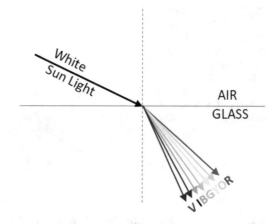

viz., violet, indigo, blue, green, yellow, orange, and red. A refraction of sunlight at the interface of two optically transparent media can spatially separate its constituent colors, a phenomenon called dispersion of light. Dispersion, as a matter of fact, owes its origin to Snell's law of refraction. Toward gaining insight into the physics of dispersion of light, we consider the refraction of sunlight on an air-glass interface, and the same has been schematically illustrated in Fig. 2.12. The speed with which light travels in a transparent medium depends on its color; red light travels slightly faster than the others, while violet is the slowest of them all. This means that the r.i. of a transparent material will be maximum for violet light and least for red light. It therefore readily follows from Snell's law that as the white light is refracted from air into the glass, its violet component will undergo maximum bending or deviation, and the red component will bend the least. Therefore, the refraction will manifest as a spatial separation of the constituent colors, traditionally denoted as VIBGYOR, of the white light as it travels into the glass. It is, however, common knowledge that a white light source, when viewed through a rectangular glass plate, does not appear to be dispersed into its spectrum of color. Notwithstanding this, dispersion indeed happens exactly the way shown in Fig. 2.12. There is a catch though. The opposite sides of a conventional glass slab are parallel, and the angle of incidence on the second interface (air-glass) for rays of each color will be exactly the same as their corresponding angle of refraction on the first interface (air-glass). The dispersion that occurred on the first surface will therefore be completely undone by the refraction occurring on the second surface. The validation of the fact captured in this figure is therefore not possible with a conventional glass slab. Newton was able to conclusively establish the dispersion of light, upon its refraction, by letting a beam of light travel through a prism. As shown in Fig. 2.13, the two sides of a glass prism are inclined to each other at an angle. In contrast to a glass slab, the second refraction will further enhance the dispersion of colors caused by the first refraction. It is of interest to note here that almost three centuries later, the concept of prismatic dispersion, originally perceived by Newton, has now been central to impart tunability in the operation of lasers (Sect. 11.3.6.3.1, Chap. 11) or purify the color of the light they give out (Sect. 6.5, Chap. 6).

Fig. 2.13 Schematic illustration of prismatic dispersion. Refraction on the first surface spatially splits the white light into its constituent colors. Refraction on the second surface enhances their spatial separation further

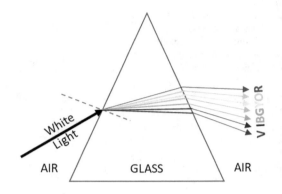

2.4 Young's Double Slit Experiment and the Wave Nature of Light

Sometime in May 2002, Stony Brook University and Brookhaven National Laboratory conducted a joint study asking physicists to nominate the physics experiments that have made the most profound impact of all time toward perceiving the design of nature. The list was published in the September issue of *Physics World*, which captured a bird's-eye view of the most imaginative physics experiments spanning approximately the last 2000 years. Thomas Young's double slit experiment with light, conducted in 1803, that conclusively reinvigorated the wave nature of light, was ranked fifth in this list.

The experimental arrangement is fairly simple and is schematically represented in Fig. 2.14. An opaque barrier with two closely spaced parallel slits etched into it is placed in front of a monochromatic[10] light source. The source is so positioned as to be equidistant from the two slits. Light passing through the slits falls on a screen kept behind the barrier. If light were made up of particles, then it would be able to leak through the two slits casting two bright stripes on the viewing screen. Instead, many alternating bright and dark stripes appear on the screen, the brightest of them being produced at a location, exactly behind the middle of the two slits, supposed to be the darkest according to Newton's corpuscular theory of light. This observation can be explained only if light travels like a wave as conceptualized originally by Huygens, and the wave theory-based interpretation is schematically illustrated in Fig. 2.15.

Let us take a small digression here to refresh our mind on the elementary knowledge of a wave. A wave can often be described as a sinusoidal propagation of some disturbance. As we know, in the case of a water wave, the disturbance is the up and down motion of water particles. It is common knowledge that when a disturbance is created at a point in a calm pool of water, a circular pattern, called water wave, spreads out from that point. Multiple such waves are seen to be formed

---

[10] Monochromaticity of a light source is a measure of the purity of the color of light it emits.

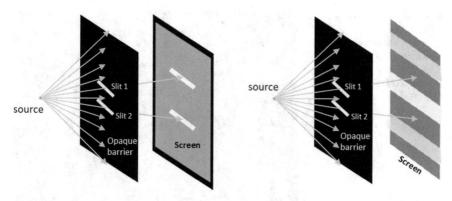

Fig. 2.14 Schematic of Young's double slit experiment with light. An opaque barrier, with two parallel slits pierced into it, is placed between the source of light and a viewing screen. The slits are symmetrically positioned with respect to the source. The source sends light all around it, but for clarity only the array of rays shining on the barrier are shown. If light were to follow Newton's corpuscular theory, then it would leak through the slits and cast two bright stripes on the screen as shown in the left picture. In reality, however, light reaches to the places on the screen forbidden by the presence of the barrier and produces alternate bright and dark zones on the screen as depicted in the picture on the right

Fig. 2.15 Wave theory-based interpretation of the formation of alternate bright and dark fringes in the Young's double slit experiment

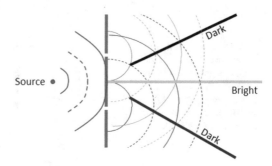

in the picture of Fig. 2.16 as raindrops fall on the water surface. Such waves, where the disturbance is perpendicular to the direction of propagation, are called transverse waves and are more abundant in nature. An exception is an acoustic wave that is longitudinal in nature. The manifestation of the oscillation of the disturbance at just one instance is shown in Fig. 2.17 for a typical transverse wave. The points of maximum and minimum disturbances are, respectively, known as crest and trough. The maximum value of the disturbance is called the amplitude (A) of the wave. If the disturbance at any point of the wave completes one full cycle of oscillation, the wave then advances through a distance called its wavelength λ. The wavelength is thus the distance between two corresponding points of the wave that are in the same phase. The number of oscillations the wave completes in a second is called its frequency (f). The velocity of the wave (v), therefore, can be expressed as

$$v = \lambda \times f \tag{2.3}$$

Fig. 2.16 Disturbances created, when the raindrops strike a surface of water, result in the formation of water waves (Credit: Patrick Nordmann, Wikimedia Commons)

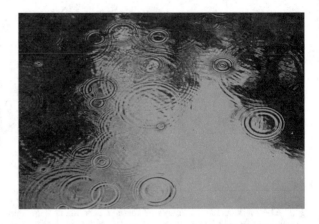

Fig. 2.17 A time frozen shot of a transverse wave. The wavelength and amplitude of the wave are normally denoted, respectively, by "λ" and "A"

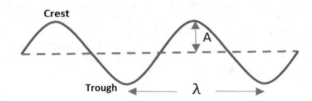

Fig. 2.18 A snapshot showing oscillatory electric (E) and magnetic (H) fields of a light wave at one instant of time. Oscillations of "E" and "H" occur on planes orthogonal to each other

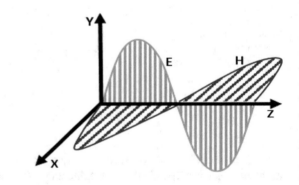

In the case of light, which is a transverse electromagnetic wave, the disturbance is the sinusoidal oscillation of both electric and magnetic fields that are always perpendicular to each other. The electric and magnetic fields shown in Fig. 2.18 are a snapshot at just one instant of time. In the case of light given out by a conventional source, the electric and magnetic fields are in continuous rotation maintaining perpendicularity between them all the time. As the polarization of light (described in a latter section in this chapter), which has a strong bearing on the operation of a laser, is determined by the orientation of the electric field, we would stick to the electric field more often in this book. Magnetic field is just as important as the electric field is, but showing just one of them at a time will add clarity to the drawings.

Let us now catch up with the double slit experiment at the point of Fig. 2.15 where we left. As the two slits are equidistant from the source, rays of light belonging to the *same wave front*[11] will strike them. The alternate solid and dotted arcs represent the wave crests and troughs, respectively. According to Huygens' wave theory of light, the two slits act as secondary sources and give out their own wavelets in unison. The wavelets emanating from the upper slit are shown in blue, while those from the lower slit are marked in red. The alternate crests and troughs are indicated as bold and dotted lines, respectively, for both the slits. The brightest stripe formed at the middle of the screen is equidistant from the slits, and therefore, the wavelets originating from the two slits will always arrive at phase, meaning that a crest will always overlap with a crest or a trough with a trough reinforcing each other, leading to the creation of the observed bright band. Dark bands, on the other hand, will be created on the locations of the screen where the crest of the wavelet arriving from one slit is lined up with a trough from the other slit or vice versa, thereby neutralizing each other. The annihilation and reinforcement of two light waves upon meeting each other is called interference of light. Deeper insight into the formation of alternate bright and dark fringes as a consequence of interference will be provided in the following section invoking the concept of path difference traveled by light from each of the two slits before they meet.

Young's double slit experiment greatly facilitated the reemergence of Huygens' wave theory, which was overshadowed for a long time by Newton's corpuscular theory, and placed optical science back on the right course. The particle theory, conceptualized by Newton in 1672, braked the flow of science for over a century, eventually culminating in the point of leading to arguably the greatest mistake in the history of physics [15] in 1818. Contesting Newton would be nothing short of a fool's errand, largely because of the great respect he earned from his peers by virtue of his momentous contributions in laying the foundation of classical physics. His authority continued unabated even beyond his death. As a matter of fact, the firm belief that light propagated as a stream of particles endured for over 75 years after his death. The corpuscular theory can be considered an early forerunner of the modern photonic concept of light and was perhaps prematurely put forward, not at the most opportune moment. Wave theory could explain almost every effect that light seemed to exhibit until of course the discovery of the photoelectric effect made in the early twentieth century. This led to the reappearance of the particle theory of light, and the realization that radiation can manifest both as particles and waves led to deeper insight into the design of nature.

---

[11] Wave front is the locus of all the points where waves starting simultaneously from the source arrive at a given instant of time. The wave front of a point source will therefore be a sphere, and the rays of light will be normal to the surface of the sphere.

Fig. 2.19 A closer look at
Young's Double Slit
experiment. The
illumination at any point on
the screen depends on the
difference of the length of
path travelled by the two
rays from the two slits up to
that point

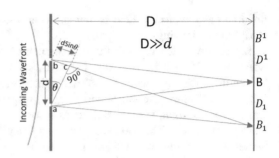

2.4.1 Double Slit Experiment and the Correlation of Wavelength of Light to Its Color

The observation of the bright and dark fringes, as we know now, can be qualitatively interpreted as the reinforcement or annihilation of the two light waves meeting on the screen. It is worthwhile at this point to translate this knowledge into an analytical expression. For this we take a closer look at an oversimplified 2D illustration of the setup of this experiment, as shown in Fig. 2.19. Upon shining by the incoming light, the two slits emit their own wavelets in unison, and only a pair of rays originating from each slit is shown here for clarity. Point B on the screen is equidistant from the slits, and therefore, the two waves that meet here will be in phase, always leading to the formation of a bright fringe at this location. Consider now the case of the next bright fringe, say the one formed at point B_1 on the screen that is located below the point B. The distances traveled by the two rays originating from the two slits before they meet at this point are different, and the difference in their path length can be equated to $d\sin\theta$ as the screen is located far away from the slits.[12] Since a bright fringe is formed at this location, the two waves that meet here must once again be in phase. The necessary condition for this to happen is.

$$d \sin \theta = \lambda; \; (\lambda \text{ is the wavelength of light}) \tag{2.4}$$

The same condition will also hold good for the bright fringe formed at the point B,[1] located on the upper side of the screen. You might probably have figured out by now that for the formation of the next pair of bright fringes, the path difference must be equal to 2λ. Let us now focus our attention on the point D_1, the location of formation of the first dark fringe. The two waves must arrive here exactly out of

---

[12] As the spacing between the two slits is infinitesimally small, the screen can be assumed to be at infinity. Two parallel rays always meet at infinity; therefore, it is implicit that rays "aB_1" and "bB_1" are parallel. Hence,

$$bB_1 - aB_1 \approx bc = d\sin\theta$$

phase, meaning that the trough of one wave meets the crest of the other or vice versa. The condition necessary for this to happen is obviously $d \sin\theta = \frac{\lambda}{2}$. The same condition will also hold true for the formation of a dark fringe at the location D'. The next pair of dark fringes will be formed when the path difference becomes $\lambda + \frac{\lambda}{2}$, i.e., $3\lambda/2$. The path difference for the formation of both types of fringes can be generalized as

$$d \sin\theta = 2N\frac{\lambda}{2} \qquad \text{for bright fringes} \qquad \text{...... 2.5}$$

$$d \sin\theta = (2N+1)\frac{\lambda}{2} \quad \text{for dark fringes} \qquad \text{..... 2.6}$$

N is an integer

The wavelength (λ), as we know, is the distance through which a wave advances as the disturbance completes one full cycle. The angle described in a full cycle is $360°$ or 2π. A path difference of "λ" is thus equivalent to a phase change of 2π. The above conditions for the formation of bright and dark fringes therefore can be expressed on the scale of phase difference (PD) as

$$PD = 0, 2\pi, 4\pi, 6\pi \ldots \ldots \text{for bright fringes} \tag{2.7}$$

$$PD = \pi, 3\pi, 5\pi, 7\pi \ldots \ldots \text{for dark fringes} \tag{2.8}$$

The above conditions for the formation of the bright and dark fringes can be readily pictured (Fig. 2.20) as the difference in the length of the paths traveled by the two waves originating from the two slits and meeting at the screen. As slit "1" is located vertically up with respect to slit "2," the wave originating from slit "2" always travels a longer distance compared to the wave originating from slit "1" for the formation of fringes above the central point of the screen. The reverse is exactly true for the formation of fringes below the central point of the screen. *With the only exception of the central bright fringe, the location on the screen for the formation of all the other bright and dark fringes is therefore dependent on the wavelength of the light source used for performing the double slit experiment.* This fact is amply demonstrated in the double slit interference pattern shown in Fig. 2.21 obtained with a white light source. As all the seven constituent colors of the white light form a bright fringe at the center, it stays white and the brightest of all the fringes. All the seven colors, as seen in this picture, are, as expected, explicit in all the higher order fringes both above and below the central bright fringe. We know that the constituent colors of the white light can be spatially separated by letting it travel through a prism. It is, therefore, possible to perform the double slit experiment with all the seven colors separately, maintaining all other conditions exactly the same. The location of all the fringes, bright or dark, with the exception of the central bright fringe, will vary with the color of the light. This means that the light of each color has a different wavelength. As an upshot of his famed double slit experiment, Young was able to make a startling revelation that red light has the longest wavelength while violet has the shortest wavelength and the remaining five lie in between. (The location of the fringes in Fig. 2.21 vis-à-vis their color bears testimony to this) The critical

Fig. 2.20 A pictorial
representation of the
condition of annihilation
(dark fringe) and
reinforcement (bright
fringe) linking the path
difference of the two waves
originating from each slit
and meeting at the screen.
Slit 1 is located vertically up
relative to slit 2

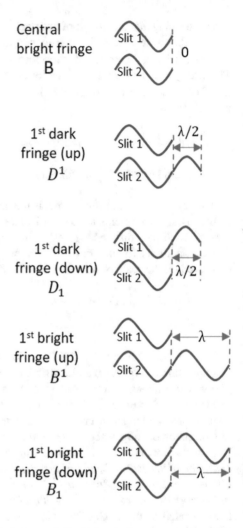

dependence of the condition of annihilation or reinforcement of two waves on their wavelengths allowed him to estimate, for the first time ever, the approximate wavelengths of each of these seven colors.

2.5 Interference of Light and Visibility of Fringes

That light added to light boosts the illumination is common knowledge, but what was not known is that light added to light can also result in darkness until Young performed his celebrated double slit experiment in 1801 to show that interference of two light waves, when out of phase, can nullify each other. Examples of interference are abundant in our everyday lives, such as the rich display of colors originating

Fig. 2.21 A double slit interference pattern formed with white sunlight (Credit: Aleksander Berdnikov, Wikimedia Commons)

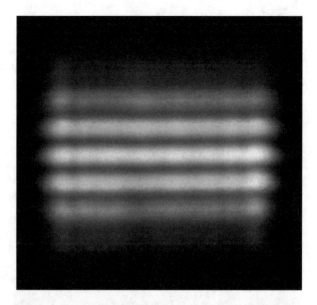

Fig. 2.22 The picture of this eyespot on the peacock's feather exhibits the Phenomenon of iridescence quite explicitly (Credit: Brocken Inaglory, Wikimedia Commons)

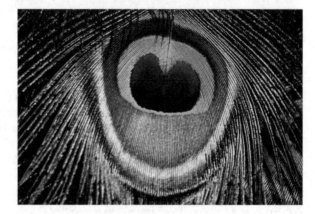

from the reflection of sunlight from an oil sleek floating on a puddle of water or from a soap bubble wandering in the air. The formation of colorful patterns of light around the Sun, Moon, and occasionally bright stars is another example. The brilliance of the iridescent[13] coloring that is known to be displayed by the eyespots on the feathers of a variety of birds (Fig. 2.22) and by several species of magnificent butterflies also owes its origin to the interference of light. The secret behind all these dazzling displays of colors started emerging once it became known that light behaves like a wave, and the physics of interference fell quickly into place.

---

[13] The phenomenon of iridescence occurs when the physical structure of an object causes light waves to interfere and create a rich array of vibrant color. The hue of the color also changes with both the angles of illumination and viewing.

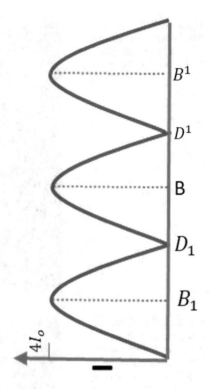

When two light waves meet together, the distribution of energy due to one wave
is modified by the presence of the other. This redistribution of light energy due to the
superposition of two light waves is called "interference of light." If these two waves
have the same amplitude or intensity, as was in the case of Young's experiment, then
they, upon superposition, will reinforce each other when in phase and annihilate each
other when in opposite phase. It is imperative at this point to take another look at the
fringe pattern formed on the screen of the double slit experiment described in
Fig. 2.19. Bright fringes are formed at the points B', B, and B_1, while dark fringes
are formed at D' and D_1 across the length of the screen as shown in Fig. 2.23.
B being equidistant from the two slits, the two waves originating from the two slits
would travel exactly the same distance in reaching this point on the screen. We also
know that the path difference of the two waves from the slits reaching either B' or B_1
is λ. The two waves that meet the screen on all three points are thus in phase boosting
thereby the intensity of light. The path difference of the two waves reaching either
D' or D_1, as we know, will be $\lambda/2$, and therefore they would nullify each other
plummeting the intensity of light at these locations to a naught. The obvious question
that arises here is how much light is cast on the screen over the intermediate regions.
For instance, as we move from D' or D_1 to B, the screen will be increasingly brighter
because the two waves now would be arriving gradually more closely in phase. On
the other hand, if we move from B' to D' or from B_1 to D_1, the screen would be
increasingly dimmer because the two waves would now be arriving with

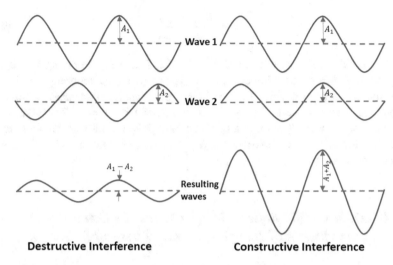

Fig. 2.24 Interference between two light waves of unequal amplitude

progressively less phase correlation. The intensity of light varies across the screen in an oscillatory manner as shown in Fig. 2.23. This thus points to the fact that by manipulating the path difference between the two interfering waves, the intensity of light upon their superposition can be varied in a controlled manner. This fact, as we shall see in the following section, can be exploited to tailor the reflectivity of a transmitting mirror, an essential prerequisite for its use as a reflector for lasers. Let us assume that each of the two waves that superimpose to interfere at every point of the screen has an intensity of value I_o. At the center of a dark fringe, where the two waves interfere destructively causing total darkness, the intensity of light dips to zero. As the total energy carried by the waves is conserved, it means that energy that disappears at one location must reappear at other locations in a manner to maintain the average intensity always at the level of $2I_o$, the intensity of the two waves putting together. This means that the intensity of light that would be cast at the center of each bright fringe would be $4I_o$.

In the case of Young's double slit experiment, the two interfering waves were of equal amplitude. Therefore, when the crest of one wave coincides with the trough of the other, they simply nullify each other to produce a complete blackness, characterized by a zero intensity of light. Consequently, the intensity of light across the screen varied between zero and a finite value. If the two interfering waves have unequal amplitudes, as illustrated in Fig. 2.24, then upon destructive interference they cannot nullify each other to produce total darkness. Consequently, the interference will manifest itself as the intensity oscillation between a maximum (I_{max}) corresponding to a bright fringe and a minimum (I_{min}) corresponding to a dark fringe. The visibility (V) of the fringes, a term first coined by Albert Michelson (1852–1931), the first American to win a Nobel Prize in Physics (and for that matter in any of the science subjects), that quantifies the contrast of the interference fringes, is defined as

$$V = \frac{I_{max} - I_{min}}{I_{max} + I_{min}} \tag{2.9}$$

Understandably, fringe with no contrast implies $I_{max} = I_{min,}$ implying $V = 0$, and fringe with perfect contrast implies $I_{max} - I_{min} \approx I_{max} + I_{min}$, implying $V = 1$.

In the case of the double slit experiment, the complete blackness of the dark fringes can be associated with $I_{min} = 0$, implying that $V = 1$. The connection between the purity of the color of a light source and the visibility of the interference fringes it forms will be described in a future chapter dealing with "Lasers in Educational Aids."

2.5.1 Tailoring the Reflectivity of a Mirror by Combining the Principle of Interference with Physics of Thin Film

A physical insight into the hue of colors and melting of one shade into another upon movement of the gaze, which we often experience around us in our everyday life (Fig. 2.25), is the key to this novel technique. This unique concept has caused a major technological breakthrough by realizing ways to manipulate reflectivity, in particular, of transmissive optics. This has eventually led to the construction of a countless variety of lasers that are in operation today. Consider, for example, the reflection of sunlight from a thin film of oil floating on water. Delving a little deeper, we realize that sunlight here undergoes reflection from basically two interfaces, viz., first between air and oil and then between oil and water. A ray diagram description of this situation is schematically represented in Fig. 2.26. A part of the sunlight is reflected from the oil-air interface, and the rest, we now know, will exhibit dispersion as it is refracted into the body of oil. Only violet and red lights are shown here for clarity. All the remaining five colors will of course lie in between. A part of all seven colors will now undergo reflection on the oil-water interface and then refraction on the oil-air interface before entering back into the air. The eye lens will now unite all

Fig. 2.25 White sunlight, upon reflection from a soap bubble, flaunts its colors by creating mesmerizing patterns of light (Credit: Umberto Salvagnin, Wikimedia Commons)

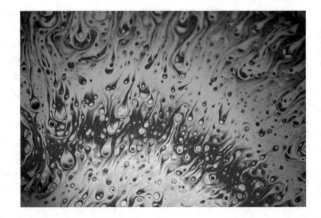

Fig. 2.26 Ray diagram indicating the physical principle of optical interference exhibited by sunlight upon reflection from the top and bottom surfaces of a thin oil film afloat on a water puddle

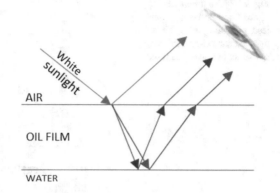

the rays onto the retina. Clearly the rays that have performed a to and fro journey through the oil traveled longer paths compared to the ones that did not. We know that two waves of same color will interfere constructively only if the distances travelled by them differ by an integral number of their wavelength. An observer will sense only this color as for all the remaining colors, the interference will be of a destructive nature. By shifting the gaze to another point on the puddle, the path difference traveled by the waves changes, and now constructive interference will occur for two waves of another color. This in brief is the origin of the spectrum of color when sunlight undergoes reflection from an oil slick floating on water.

A similar concept can be made use of to tailor the reflectivity of an optical element of transmissive nature, called partially transmitting mirror, which is mandatory for coupling the beam of light out of the laser. The situation is less complex here as we focus on the light of a single color and normal angle of incidence. Thin films of oil, however, must be replaced by another dielectric material that allows the transmission of light of a specific color. Glass is known to offer transmission over a wide spectral range spanning from near-infrared through the visible into the vacuum ultraviolet regions and therefore is a preferred optics material for a wide variety of lasers. Only approximately 4% of the light, as we know, when incident normally on an air-glass interface is reflected back into the air. The reflectivity needs to be appropriately scaled up to make it suitable for use as a laser mirror. For this, a thin film of a dielectric[14] material, transparent to the wavelength of the laser, is deposited on the side of the glass substrate that would face the active medium of the laser (Fig. 2.27). The light wave falling normally on the mirror will be reflected back from both the outer and the inner surfaces of the dielectric film and interfere with each other, much the same way as in the case of the oil film floating on water. The wave reflected off the inner surface will travel an additional distance equal to twice the thickness of the film compared to the wave reflected from its front face. Constructive interference between these two reflected waves will occur if the distances traveled by them differ

---

[14] A dielectric is a nonmetallic substance that either is a poor conductor of electricity or doesn't conduct at all. Unlike metals, they have no free or loosely bound electrons.

Fig. 2.27 An amplified
view of one end of a typical
laser indicating the details of
the dielectric coated mirror
to extract the laser beam out

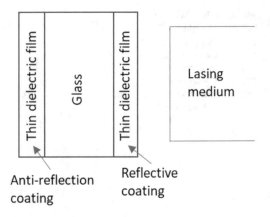

by an integral number of the wavelengths. If the thickness of the dielectric film is such that constructive interference does indeed occur between these two reflected waves, no light will be left to travel into the glass. It is equivalent to saying that the mirror in this case will ideally offer a 100% reflection. As we would learn in a future chapter, although such a mirror may perform as an end mirror of a laser, it would not work as an output mirror, as no light can leak through it. A partially reflective mirror, which would offer a fraction of light to travel through, will fit the bill as an output mirror. We have learned in the previous section that by manipulating the path difference between the two interfering waves, the intensity of light, upon their superposition, can be varied in a controlled manner. (Take another look at the variation of intensity across the interference fringes as depicted in Fig. 2.23.) The techniques of thin film deposition are vastly matured and offer absolute control on the thickness of the dielectric film. This would allow manipulation of the degree of constructive interference between two light waves and, in turn, the reflectivity offered by the film. This, however, does not solve the problem completely though. For instance, an 80% reflective film will allow 20% of the light shining onto it to travel through into the glass (Fig. 2.28). Upon reaching the far end of the glass, 4% of this light will be reflected back from the glass to air interface unless a second dielectric film is coated on this end to prevent the occurrence of this reflection. The thickness of the film now has to be so chosen as to ensure a complete destructive interference between the light waves reflected from its inner and outer surfaces. Termed as an antireflection coating, this dielectric film will allow emergence of the entire 20% of the light that could leak through the other dielectric film. A glass plate with antireflection film coated on both ends will act as a transmitting window and is often used to vacuum seal the two ends of a gas laser. This is a rather qualitative insight into the physics of manipulating the reflectivity of transmissive optics to make it suitable for use as an optical element in the operation of lasers. In reality, however, the manipulation issue is far more complex often requiring judicious combination of multiple dielectric layers and their refractive indices. Multilayer-based AR coatings, with the ability to eliminate Fresnel reflection across the entire visible spectrum, find wide application as reflection arresters in spectacles, telescopes and microscope optics, photo-frames, etc.

Fig. 2.28 If a partially reflective coating is affixed on one end of the glass mirror and the other end is open to air, then a part of the light is reflected back from the air-glass interface (top trace). The loss of light through such reflection can be avoided through insertion of a second dielectric film on the other end that works like an antireflection coating (bottom trace)

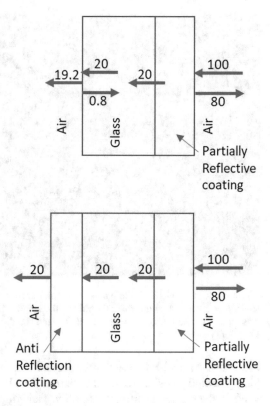

2.6 Diffraction of Light

The diffraction of light is the process of its bending or spreading out upon passing through an aperture or interacting with an obstacle of size comparable to its wavelength. Diffraction thus makes possible for light waves to reach the geometrical shadow of the object. If the size of the apertures or the obstacles far exceeds the wavelength of light, then its passage will be unhindered by their presence. An analogy can be drawn when you walk through a gate. If the gate is narrow, then you need to judiciously maneuver yourself like taking a $90°$ turn or whatever to squeeze through. If, on the other hand, the gate is wide, your movement will be almost oblivious to its presence. It is common knowledge that the phenomenon of diffraction is exhibited not only by light but also by all other waves when they encounter obstacles of size closer to their respective wavelengths. For example, water waves upon hitting a boat bend around it, ocean waves can bend around a jetty,

Fig. 2.29 The exquisite multiple shades of color seen in this picture are generated when white sunlight is diffracted by the water droplets in the cloud. This magnificent cloud iridescence effect was captured at the sunrise over the gulf of Thailand (Credit: Phuket@photographer.net; Wikimedia Commons)

and sound waves bend around more common objects allowing us to hear someone hiding behind a wall or some other barrier. The longer the wavelength of the sound wave is, the greater its diffractive ability. Many forest-dwelling birds take advantage of this to communicate over longer distances. The long wavelength hooting sound of owls, for instance, can easily diffract around forest trees and carry much further than the short wavelength tweets of a wide variety of songbirds. Owl's hoots, which can be heard over several miles, help them to attract or communicate with a mate or to locate a stranded owlet. X-rays, microwaves, and radio waves also exhibit diffraction under suitable conditions, and matter waves are also no exception. Francesco Maria Grimaldi (1618–1663), an Italian polymath, who coined the term diffraction, is credited with the discovery of this interesting behavior of light. The effects arising due to diffraction of a variety of lights are frequently seen in our everyday life. For instance, light from celestial bodies upon diffraction by tiny particles present in the atmosphere results in the appearance of bright rings around the Sun, the Moon, or other bright sources. The speckle pattern produced when an optically rough surface is shone by light emitted from a laser is also a phenomenon of diffraction. Laser diffraction patterns are often used to create jaw-dropping displays in laser shows. The most glaring example, however, is the dazzling color thrown onto the sky when sunlight trickles through the cloud after sunrise or before sunset (Fig. 2.29).

You may be intrigued as to why diffraction occurs only when the size of the obstacle/aperture is comparable to the wavelength of the light. A quantum-classical

Fig. 2.30 Plane wave
shining on a slit of width
Δx. The wave spreads
through an angle $\Delta\theta$ after
diffraction

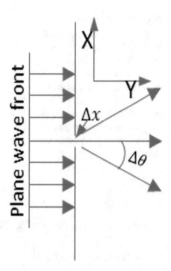

connection, by invoking Heisenberg's uncertainty principle,[15] offers a qualitative
explanation for this. To this end, we consider a simple example of shining light onto
a slit of width Δx. Light would pass straight through the slit with an angular spread of
$\Delta\theta$ as shown in Fig. 2.30. The momentum p associated with light of wavelength λ is
given by.

$$p = \frac{h}{\lambda} \quad (h \text{ is Planck's constant}) \tag{2.10}$$

The momentum lies entirely along the Y direction before the passage of the wave
through the slit with no restriction imposed on the wave in the X direction. As the
wave passes through the narrow slit of size Δx, it experiences a restriction from the
X direction. This understandably sets in a positional uncertainty equal to the width
Δx of the slit to the wave in this direction. As a result, there appears a corresponding
spread (Δp_x) of the momentum of the wave in the X direction, the minimum value of
which is governed by the uncertainty principle as

$$\Delta p_x \times \Delta x \geq \frac{h}{4\pi} \tag{2.11}$$

---

[15] Heisenberg's uncertainty principle: There is an inherent limit imposed by the nature on the
combined accuracy with which a pair of certain canonically conjugate variables like position and
momentum or energy and time can be determined. If effort is expended to measure one variable
more accurately at one instant, the other variable becomes intrinsically more unpredictable at that
instant. If Δx and Δy are the uncertainties in the measurement of two such variables x and y, then
uncertainty principle can be expressed mathematically as $\Delta x \, \Delta y \geq h/4\pi$, where h is the Planck's
constant.

Fig. 2.31 Schematic illustration of the single slit diffraction of light. When the width of the slit is comparable to the wavelength, the diffraction manifests itself as a spatial oscillation of intensity (I) as recorded on a screen placed long distance away from the screen

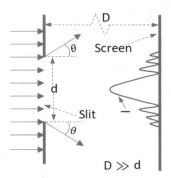

As $p\Delta\theta \approx \Delta p_x$ for a small spread of momentum,

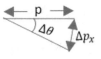

we therefore obtain

$$\Delta\theta \geq \frac{h}{4\pi p\Delta x}$$

i.e., $\boldsymbol{\Delta\theta} \geq \dfrac{\lambda}{4\pi\Delta x}$, substituting λ for h/p (2.12)

This equation provides illuminating insight into the phenomenon of diffraction of light. If the slit is much too long compared to the wavelength, i.e., if $\Delta x \gg \lambda$, then $\Delta\theta \sim 0$ meaning that the angular spread of the wave upon leaving the slit will be vanishingly small. This is equivalent to saying that light will pass through a large aperture unhindered. If the slit width closely matches the wavelength, i.e., if $\Delta x \approx \lambda$, then the angular spread of the wave after its passage through the slit becomes appreciable. Finally, if the width of the slit is much less than the wavelength, i.e., if $\Delta x \ll \lambda$, the light understandably spreads out over the entire area on the other side of the slit. Quantum uncertainty arising from the restriction imposed on the wave by the aperture therefore qualitatively describes the manifestation of the diffraction of light through apertures of varying sizes.

2.6.1 Single Slit Diffraction of Light

As we know now, a beam of light, upon passing through a slit, will experience an angular spread of "θ" if its wavelength is close to or more than the width of the slit. This angular spread manifests as a spatial oscillation of intensity of light on a screen placed in front of the slit but at a long distance away from it. The light cast on the screen is qualitatively similar to what is seen in the case of Young's double slit experiment. There is a significantly bright central zone and on either side, a series of alternate bright and dark bands appear (Fig. 2.31). The intensity of the bright fringes, however, progressively drops. Joseph Ritter Von Fraunhofer (1787–1826), a German physicist, is credited as being the first to offer an insight into the physics of the single slit diffraction pattern. We provide below a qualitative, chiefly graphic, explanation of the far-field[16] diffraction pattern based primarily on the principle of interference of light. It is worthwhile to mention here a few facts that are central to understanding the mechanism behind the formation of far-field single slit diffraction patterns:

- According to the Huygens principle, all the rays originating from all the different points of the slit after diffraction are in phase.
- The angular spread of the beam of light through "θ" caused by diffraction would ensure the presence of a group of parallel rays travelling outwardly in all possible directions ranging from 0 to θ.
- Parallel rays of light starting from the tiny slit of size close to the wavelength of light and heading toward a common destination meet at the far field.

The situation after the beam of light undergoes diffraction has been captured in a set of ray diagrams constituting Fig. 2.32. Consider the case represented in trace (a) showing a group of parallel rays that move forward without any deviation after the diffraction, meaning their angle of diffraction is zero. All these rays are in phase and head straight toward the central location of the screen. They understandably would travel equal distances to reach the screen and therefore will still remain in phase. These rays will therefore constructively interfere leading to the formation of the central bright fringe on the screen. Consider now the situation described in trace (b) depicting a second group of parallel rays that have been diffracted through an angle of θ_1 in the upward direction and accordingly head toward a corresponding common location on the screen. Understandably in order to reach the screen, each ray will now travel a different distance and therefore will arrive in or out of phase. For instance, the ray from the bottom most point of the slit in this particular case travels exactly one wavelength λ farther than the ray at the top. The ray at the center will travel $\lambda/2$ distance farther compared to the ray at the top and thus would arrive at the screen exactly out of phase and would nullify each other. Likewise, two more rays one slightly below the top of the slit and the other from slightly below its center will also arrive exactly out of phase at the screen and cancel each other out. In fact, each ray originating from the slit will have a corresponding ray that interferes destructively, meaning that a reduction in intensity will occur at this angle. This will lead to the formation of the first dark zone above the central bright spot. The group of parallel rays diffracted through θ_1 in the downward direction (trace c) will likewise result in the formation of the first dark zone but now below the central bright spot. The condition for the formation of the first dark fringe on either side thus can be mathematically expressed as

---

[16]Far field in a scientific parlance refers to a location so far away from the point of occurrence of a scientific event that its effect can be considered to be independent of distance beyond this point.

Fig. 2.32 Light passing through a single slit undergoes angular spread and may interfere constructively or destructively depending on the angle at which the rays arrive on the screen. "Trace a" shows the array of parallel rays that emerge un-deviated after diffraction and constructively interfere to cast a bright fringe at the center of the screen placed at the far field. Trace b and c show the parallel rays that cast the first dark spots above and below the central bright zone, respectively. Trace d and e illustrate generation of first bright fringes above and below the central spot, respectively

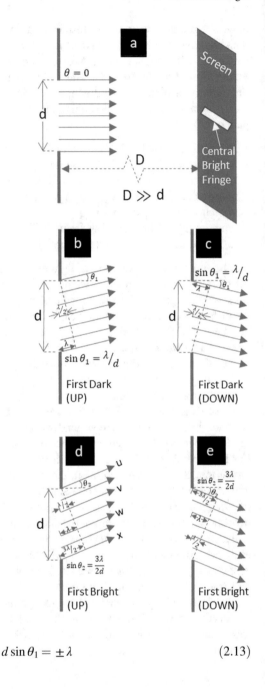

$$d \sin \theta_1 = \pm \lambda \qquad (2.13)$$

To understand the formation of the first bright fringe above the central bright spot, we consider the group of parallel rays (trace d) that are diffracted upward through an angle θ_2. For this angle, the length of paths that the rays originating from the bottom and top of the slit travel to reach the screen differs by $3\lambda/2$. For easier comprehension, a few specific rays here have been labeled. Ray w will travel exactly a wavelength longer than ray u and thus will reinforce each other on the screen. In fact, all the rays lying between u and v will arrive at the screen in phase with a corresponding ray lying within w and x. The rays between v and w will, however, miss out on the opportunity of such a phase pairing. This, in turn, results in the formation of a bright fringe here but not as intense as the central maximum. Trace e indicates the group of parallel rays diffracted through an angle θ_2 downward to cast the first bright fringe below the central maximum. The formation of the higher order alternate dark and bright zones, the characteristic features of single slit diffraction, can be understood by considering the group of parallel rays diffracted through correspondingly higher angles.

In a future chapter, we shall know that a multiple number of single slits, when placed in a periodic manner to provide alternate transmission and opaque zones, serve as an excellent tool for spatially separating the colors of the input light. This arrayed structure of slits is called a diffraction grating that finds wide application as a dispersive element in a variety of optical instruments such as monochromators and spectrometers. We shall also learn how a grating can be made to serve as a tuning element in the operation of a laser. In fact, the grating is an integral part of almost all kinds and varieties of tunable lasers that are in operation today.

2.7   Scattering of Light

Scattering is a phenomenon wherein light, upon interaction with a particle, microscopic or macroscopic, is redirected in many different directions. Let us take a small digression here and refer to the famed picture called Earthrise, a snap of Earth taken from the lunar orbiter during the Apollo 8 mission, the first crewed voyage to orbit the Moon. This allowed humankind, for the first time ever, to view the sky through the lens of a camera located in the vicinity of a natural celestial body where there is no medium to scatter sunlight. We replicate here the Earthrise (Fig. 2.33), acclaimed as the most influential environmental photograph ever taken [16], and seek to decode the science this picture holds. The Sun, hiding on the other side of the camera, is obviously the source of light for this picture. The lunar surface at the foreground and the Earth at a distance are seen basically by sunlight reflected off them. Sunlight traverses through the empty space between the Moon and the Earth, and in the absence of any matter to return light back to the camera lens, it appears pitch dark. Even the space all around the Earth, which we perceive as our blue sky, appears no less dark from the Moon! The only difference being that we look at the sky through a layer of atmosphere that surrounds the Earth while the Moon has no atmosphere. Unmistakably then, our sky would be just as black as it is in this picture if the Earth didn't have an atmosphere. It is also apt in this context to take a look at the mesmerizing picture (Fig. 2.34) of the early morning sunlit Martian sky captured by NASA's insight lander during its mission to the red planet in 2019. The bright

Fig. 2.33 The rise of Earth as seen from near the Moon. This legendary picture, shot on December 24, 1968, during Apollo 8 lunar mission, is credited to astronaut William Anders (Source: NASA, Wikimedia Commons)

Fig. 2.34 This picture showing a slice of the blue sky above the Domoni Crater of Mars was captured by the Instrument Deployment Camera (IDC) fitted on the robotic arm of NASA's insight lander during its Martian mission. The Sun too is visible, albeit dimly, in the image (Source: NASA, Credit: Kevin Gill, Wikimedia Commons)

sky, in total contrast to the Earthrise photo, bears signature of the presence of atmosphere in Mars to scatter sunlight to reach the camera.

Let us dig a little deeper to determine the role the atmosphere, be it of Earth or Mars, plays in this context. Our atmosphere, for instance, is a layer of gases, called air, surrounding the Earth that has been retained by gravity. Air, as we know, is predominantly made up of nitrogen and oxygen molecules. In 1871, Lord Rayleigh (1842–1919), a British physicist, provided a theoretical description of the scattering of light involving particles smaller than the wavelength of the light. According to this theory, now popular as Rayleigh scattering, which paved the way for his securing the 1904 Nobel Prize in Physics, the particle can scatter a part of the incident light uniformly all around it. The intensity of the scattered light (I_s) is inversely proportional to the fourth power of wavelength of the incident light, i.e.,

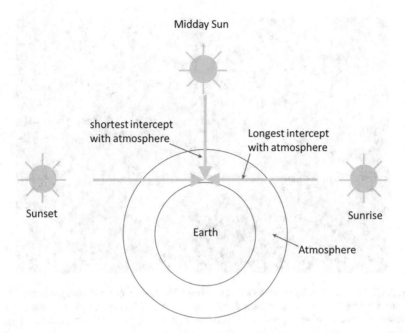

Fig. 2.35 The intercept of the sunrays with the atmosphere is maximum at sunset or sunrise and minimum at midday

$$I_s \propto \lambda^{-4} \tag{2.14}$$

The white sunlight is made up of seven colors of which violet possesses the smallest wavelength and red the largest and the remaining five lie in between. The size of both nitrogen and oxygen molecules, the major constituents of air, is less than even the wavelength of violet light. Clearly, both of these molecules, upon being shone with sunlight, will scatter all seven colors of white light following the description of Rayleigh, meaning that the violet color will be scattered the most and the red the least. However, there is not as much violet or indigo in the sunlight as there is blue. By the time the sunray reaches the Earth's surface, after traversing the entire atmospheric column, it would have lost a large amount of blue through Rayleigh scattering. Finally, then, as we look away from the Sun into the sky, it is more likely that we would catch the scattered blue light making the sky appear blue. The bluish hue of the Martian sky, as evident from Fig. 2.34, also testifies the invariance of Rayleigh's theory. The moon has no atmosphere, so there is nothing to scatter the sunlight, which is why the entire sky in the Earthrise photo appeared absolutely dark.

Red color of the Sun at sunrise or sunset also owes its origin to the phenomenon of Rayleigh scattering. As is seen in Fig. 2.35, the length of atmosphere traversed by the sunrays is maximum both at sunrise and sunset and minimum at midday when the Sun is directly overhead. The longest intercept with the atmosphere allows sunrays to shed almost all the colors, through Rayleigh scattering, retaining only a

Fig. 2.36 The Sun, profoundly subdued at the time of its setting, appears remarkably red. In this picture, the Doi Suthep temple, a sacred site to the people of Thailand, is seen to make a stark silhouette against the dimming red sun (Credit: Mi. Suthat Fongmoon, Wikimedia commons)

shred of red. At sunset or sunrise, the Sun, therefore, not only turns incredibly red (Fig. 2.36) but also can be viewed directly by the naked eye.

It is common knowledge that on a clear and cloudless bright day, particularly following a shower, the bluish hue of the sky appears much deeper. The slice of the fascinatingly deep blue sky on a clear and bright sunny morning following a brief but heavy downpour captured in Fig. 2.37 bears testimony to this fact. The rain basically cleans up the air by settling down the dust and other suspended particles from it. The obvious effect of the presence of these impurities, of size far exceeding the light wavelengths, in the air is, therefore, to trim the blue of the sky a shade or two. You may now begin to wonder if Raleigh scattering makes the sky appear blue, what is it that makes the cloud floating in the same sky often appear white? The cloud looks white as the white sunlight gets scattered from the water droplets that the cloud is made up of, preserving all its constituent colors. Similar to dust and impurity particulates, water droplets are also much larger than air molecules and do not seem to follow the Rayleigh scattering formula. In 1908, Gustav Mie (1868–1957), a German physicist, presented a description of the scattering of light by particulates of sizes exceeding its wavelength. Qualitatively speaking, the intensity of scattering, known as Mie scattering which has no restriction on the upper size of the particulates, is invariant of the wavelength of light. This thus readily explains why the presence of particulates in the atmosphere reduces the bluish hue of the sky or for that matter why clouds often appear white.

Note that for both Rayleigh and Mie scattering, the light photons do not deliver any energy to the scatterer and are essentially redirected in all possible directions with their energy intact. Such scattering processes without involving any exchange

Fig. 2.37 This photograph taken by me on a bright sunny morning in the San Francisco Bay Area following a shower reveals the captivating blue hue of our sky

of energy between light and the scatterers are called elastic scattering. The incidences where photons can both gain and lose energy to the scatterer are not rare in physics and are termed inelastic scattering of light. A few notable examples are Raman scattering, Brillouin scattering, and Compton scattering. The advent of lasers has added multiple dimensions to all these effects, opening up new frontiers of physics and mind-boggling applications, unthinkable had laser not existed, and will be covered later in the book (V-II).

2.8   Polarization of Light

A majority of readers will have either used polarized sunglasses at one time or another or are aware of their functionality. The polarized glasses have a special ability to make the viewing unbelievably soothing by way of cutting the glare of light originating from a bright source or light reflected off a shiny object. The efficaciousness of a polarizer in removing sun glare is illustrated in the pictures of Fig. 2.38. These are often used while driving in the night or snow gazing on a bright sunny day. To understand the physics behind the ability of these special glasses to slash the glare of light, we need to take a closer look at the transverse nature of light waves. Light, as we know now, is a transverse wave wherein the disturbance is perpendicular to the direction of propagation. The disturbance is the sinusoidal oscillation of both electric and magnetic fields that are again perpendicular to each

Fig. 2.38 This picture clearly illustrates as to how a sunlit object would look when viewed through a polarized glass (left) and naked eye (right). The polarizer cuts off the Sun's glare originating from the water ripples allowing a clear and unobstructed view of the river bed (Credit: Dmitry Makeev, Wikimedia Commons)

Fig. 2.39 The electric (E) and magnetic (H) fields oscillate in a plane transverse to the direction of propagation (Z) of light. For the three light waves shown here, although E and H are oscillating in different directions, they are always perpendicular to each other

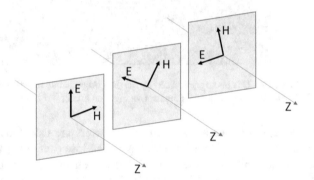

other. There is obviously no restriction placed on the direction in which electric and magnetic fields can oscillate as long as they are orthogonal to each other. This fact has been illustrated in the traces of Fig. 2.39 where three waves have been captured with rotating electric and magnetic fields with their orthogonality intact. A ray of light is made up of multiple waves. For a conventional source of light, in each of these waves, the direction of oscillation of the electric or magnetic fields will vary randomly. Considering only the electric field, for clarity, as a representation of a wave, a light ray is schematically depicted in Fig. 2.40. Light emitted by a conventional source, such as the Sun, therefore, has an electric field oscillating randomly in all possible directions and is called randomly polarized or unpolarized light. A polarized light understandably has an electric field oscillating only in a particular direction. In the case of a horizontally polarized light, denoted normally as *p* polarization, oscillation of the electric field always occurs on a horizontal plane as depicted in Fig. 2.41. For vertically polarized light, denoted as *s* polarization, the oscillation of the electric field, on the other hand, will stay confined in a vertical plane (Fig. 2.42). Light with random polarization will essentially have an electric field oscillating on all the planes lying between 0 and 2π angles. Electric field being a vector, the randomly occurring fields can be resolved into two components, one

Fig. 2.40 Schematic illustration of the random orientations of the electric field of the waves constituting a ray of unpolarized light

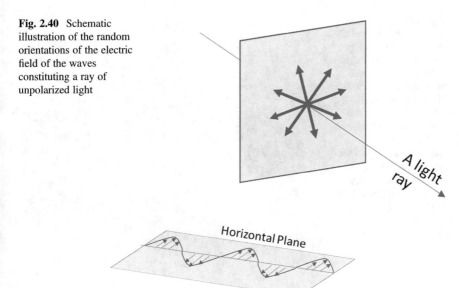

Fig. 2.41 In case of a horizontally polarized light, the oscillation of the electric field, indicated by the arrows, is always contained in a horizontal plane

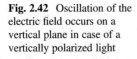

Fig. 2.42 Oscillation of the electric field occurs on a vertical plane in case of a vertically polarized light

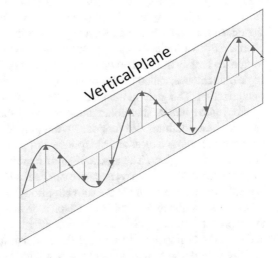

lying on the horizontal plane and the other on the vertical plane. It is, therefore, equivalent to saying that unpolarized light comprises horizontal and vertical polarizations in equal amounts.

Having acquired this elementary knowledge pertaining to the polarization of light, we are now in a position to figure out as to how a polaroid glass eliminates the glare arising from the reflection of light. In 1811, a Scottish physicist, by the name David Brewster (1781–1868), made a profound discovery in the area of optics

Fig. 2.43 A transverse wave set on a vertical plane in a rope is blocked by a horizontal slit (**a**) and moves without any hindrance through a vertical slit (**b**). The element of surreality, readily apparent in this illustration, is deliberate. This is to allow an unobstructed view of the wavy rope and the imaginary plane of its vibration, not possible in practice

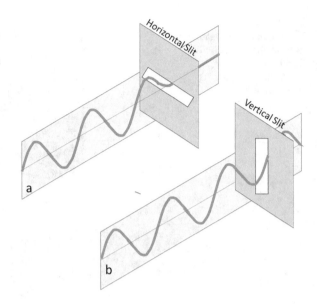

that is now popular in the scientific world as the Brewster or polarization angle. This discovery has a strong bearing in many branches of science and, as will be shown in a latter chapter, is a key to obtaining polarized output from a laser in the simplest and most inexpensive way wherein practically no light is lost into the unwanted polarization. In summary, Brewster found that the reflection of unpolarized light from the interface of two optically transparent media depends on the angle of incidence and comprises largely the vertical or s component of polarization. In fact, for a particular angle of incidence, called the Brewster angle, reflected light comprises entirely of the s polarization. The glare in the vision is thus primarily linked to this polarization. A polaroid glass blocks this polarization to make the uncompromised viewing both pleasing and soothing. Professional photographers often make use of the same principle to acquire a clear and sharp image of an object beneath a reflective surface such as water or glass. The mechanism of removal of waves of a particular polarization in the case of a transverse wave can be readily understood by considering, for example, the case of such a wave moving through a rope. In the first experiment, we set in an oscillation at one end of a rope on a vertical plane and make the other end of the rope pass through a horizontal slit (Fig. 2.43a). The horizontal slit would understandably impede any vertical disturbance, and the wave would simply die out. We now repeat the experiment but now slide a vertical slit through the other end of the rope (Fig. 2.43b). The vertical wave now continues its motion without any hindrance through the vertical slit. Likewise, an oscillation in the rope, when impressed on a horizontal plane, will travel through a horizontal slit unhindered but will be blocked by a vertical slit. Analogous to such selections of the polarization of rope waves through the rotation of a slit, the polarization of a beam of light can

Fig. 2.44 Schematic of the experimental setup to monitor the reflection of light from an air-glass interface as a function of angle incidence for both s and p polarizations

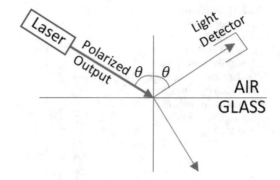

also be selected by letting it pass through a light polarizer[17] and rotating it appropriately. In the following section, we would also learn as to how the property of the Brewster angle can be exploited to convert an unpolarized beam of light into a linearly polarized beam.

2.8.1 Brewster Angle and Linearly Polarized Light

We know that when light shines on the interface between two optically transparent media, a part of it is reflected back into the incident medium, while the rest is refracted into the second medium. The realization that the amount of reflection of light depends not only on the angle of incidence but also on its state of polarization was instrumental for the conceptualization of Brewster's law. To gain deeper insight, it is imperative to study the reflection of light from an interface such as that of air-glass as a function of the angle of incidence for both p and s polarized light. An ideal light source for performing this experiment will be a laser capable of emitting light with p or s polarization. The mechanism of generating linearly polarized light from a laser will be described in a future chapter (Sect. 7.5.1.2.1, Chap. 7). The emission from the laser, alternately in p and s polarizations, is made to shine on an air-glass interface, and a light detector is located to capture the reflection and measure its power as a function of the angle of incidence θ. The setup of the experiment is schematically shown in Fig. 2.44. The functional dependence of reflectivity on the angle of incidence that has a strong bearing on the optical science in general is recorded in Fig. 2.45 for both p and s polarized lights. As seen, there is a notable contrast in the behavior of the s and p polarizations. At a normal angle of incidence ($\theta = 0°$), the reflectivity for the air to glass interface is ~4% for both polarizations. In the case of s polarization, the reflectivity exhibits a rise with

---

[17]A polarizer is an optical element that at a particular orientation allows light of a specific polarization to pass through while blocking the other polarization. Imparting a 90° rotation to the polarizer will flip the polarization of the passing light.

Fig. 2.45 Percentile reflection of light off an air-glass interface as a function of angle of incidence for both "p" and "s" polarizations. The angle of incidence θ_B at which reflection of "p" polarized light disappears is called Brewster's angle

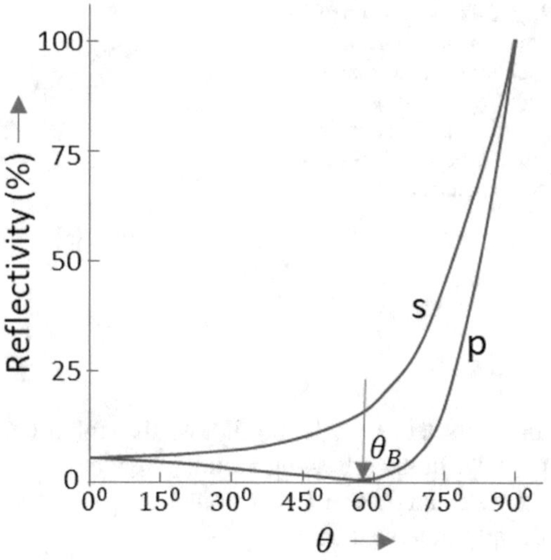

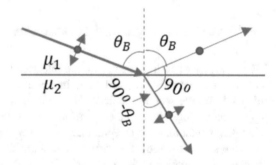

Fig. 2.46 When light is incident at the Brewster's angle (θ_B) on the interface of two optically transparent media, it is refracted into the second medium with an angle 90°-θB. μ_1 and μ_2 are the refractive indices of the two media

increasing θ, initially slowly and then rapidly in the vicinity of the grazing angle of incidence ($\theta = 90°$). The reflectivity of p polarized light behaves in a contrasting manner gradually dropping first to zero at $\theta \approx 57°$ and then exhibiting a rapid rise as θ approaches 90°. The angle of incidence θ_B, at which the reflectivity of p polarized light completely disappears, is called Brewster's angle. As would be seen, approximately 15% of the s polarized light is reflected at this angle of incidence. Additionally remarkable is the fact that at the Brewster's angle of incidence, the reflected and refracted rays make exactly a right angle between them.[18] This makes the application of Snell's law for refraction quite straightforward, paving the way to derive an expression for Brewster's angle. For this, we consider the case of a ray of unpolarized light striking the interface of two media of refractive indices μ_1 and μ_2 at the Brewster's angle of incidence, θ_B (Fig. 2.46). The reflected light will understandably be devoid of any p polarization making the refracted light more enriched in this

---

[18] The orthogonality of reflected and refracted rays at the Brewster's angle of incidence forbids reflection of light with p polarization as it violates the transverse nature of light.

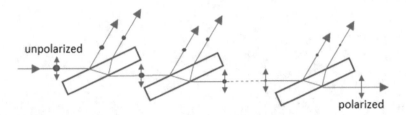

Fig. 2.47 At Brewster angle of incidence, 15% of the "s" polarization of the unpolarized light is reflected off at each interface (air to glass or glass to air) gradually enriching the "p" polarization content of the refracted ray. Stacking a large number of Brewster plates on its path, the unpolarized light can, therefore, be converted into "p" polarized light

polarization. The orthogonality of reflected and refracted rays clearly renders a value of $(90^\circ - \theta_B)$ to the angle of refraction. Application of Snell's law, therefore, leads to

$$\mu_1 \sin \theta_B = \mu_2 \sin \left(90^\circ - \theta_B\right)$$

$$\Rightarrow \tan \theta_B = \frac{\mu_2}{\mu_1.}$$

$$\Rightarrow \theta_B = \tan^{-1} \frac{\mu_2}{\mu_1} \qquad (2.15)$$

As we shall see in a latter chapter, the above equation, connecting Brewster's angle to the refractive indices of the two media defining the interface, is central to obtaining polarized light from a laser. For an air-glass interface, $\mu_1 \approx 1$ and $\mu_2 \approx 1.5$ yielding $\theta_B \approx 57^\circ$. This precisely is the experimentally determined value of Brewster's angle for the air-glass interface. If a randomly polarized beam of light is made to shine on a glass plate, immersed in air, at an angle of incidence of 57°, then ~15% of its content of s polarization will be reflected off at each interface (Fig. 2.47). Thus, the refracted beam of light is depleted in s polarization and, in turn, becomes enriched gradually in p polarization as it passes through such plates successively. A linear polarizer can thus be formed by stacking a large number of such Brewster plates one after another. In contrast, a lone Brewster plate, as we shall find afterward, suffices to derive polarized output from a laser, such is the dominance of stimulated emission!

Chapter 3
Quantization of Energy

3.1 Introduction

Until the discovery of the photoelectric effect in 1902 [1], everything seemed to be falling in place and physics moved forward with no qualms. The behavior of light and matter, the two facets of nature, could be accurately described by the wave theory and Newtonian mechanics, respectively. The discovery of the photoelectric effect, which unequivocally established the dual nature of light, viz., its ability to manifest both as a particle and wave, turned things around forever. It was not long before the dual nature of waves led to the realization of the dual nature of particles as well. This revolutionary insight into the behavior of particles, although it fetched a 1924 Nobel Prize in Physics to Louis de Broglie, dragged physics into rough weather. It is common knowledge that waves, when bounded by a constraint, can exist only at discrete frequencies or wavelengths. For example, in the case of a rope, rigidly clamped at the two ends, only waves, half of whose wavelength times an integer equals the rope's length, can be impressed upon it. Alternatively, an optical cavity can contain only that light whose wavelength also bears a similar relationship to the length of the cavity. Ascription of the wave nature to a particle, therefore, means that if the particle is confined within a region in space, the wavelengths that fit into this space will only be allowed. This restriction on wavelength, in turn, restricts the values of energy [17] that the particle can possess. While quantization of matter into lumps of atoms, molecules, and the like was well known, the very concept of discretization of its energy states seemed to have flummoxed the physics community in the early part of twentieth century. After all, quantization of energy of a particle would be analogous to having only certain specific speeds at which a car can travel, a soccer ball can be kicked, or a baseball can be struck! Fortunately, matter does not behave this way in our world despite its wavy nature. Where lies the catch then? This chapter has been planned to strike a sense of understanding, albeit qualitatively, to the mind of the readers on the implication of quantization of the energy states in the

© The Author(s), under exclusive license to Springer Nature Switzerland AG 2023   59
D. J. Biswas, *A Beginner's Guide to Lasers and Their Applications, Part 1*,
Undergraduate Lecture Notes in Physics,
https://doi.org/10.1007/978-3-031-24330-1_3

Fig. 3.1 Discretization of
the de Broglie wavelengths
of a particle trapped in a box

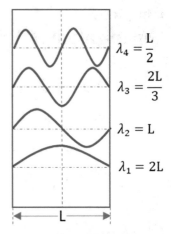

$$\lambda_4 = \frac{L}{2}$$

$$\lambda_3 = \frac{2L}{3}$$

$$\lambda_2 = L$$

$$\lambda_1 = 2L$$

macroscopic world we are familiar with. The chapter also mirrors the manifestation
of the wave nature of a bound microscopic particle into the spectral emission from
atoms and molecules that, in turn, would eventually lead to the realization of lasers.

3.2   A Bound Particle and Discretization of Its Energy

To gain deeper insight into the wavy matter, it is imperative that we consider a
particle of mass m that is trapped inside a box and bouncing back and forth between
its two walls spaced by a distance of L. We further assume that these two walls are
infinitely rigid and the particle, therefore, cannot lose any energy on the walls every
time it hits them. The situation is schematically illustrated in the traces of Fig. 3.1.
Considering the wave picture, the wave length λ of this particle, which is analogous
to the particle nature of light, called the de Broglie wave length, can be shown to
be [18].

$$\lambda = h/mv \tag{3.1}$$

where h is Planck's constant and v is the velocity of the particle. Drawing an analogy
to the waves set in a rope with its two ends rigidly clamped, we surmise that here too
an infinite number of waves can be associated with the particle all of which will be
characterized by zero displacements at the two rigid walls. It thus readily follows that
the de Broglie wavelengths are intricately linked to the spacing L between the walls
and the longest of them λ_1, which also happens to be the fundamental of all these
possible waves, will be obviously given by

$$\lambda_1 = 2L$$

And the next one will be $\lambda_2 = L$, and the next $\lambda_3 = \frac{2L}{3}$, the next $\lambda_4 = L/2$, and so on.

Upon generalizing all these into a single formula of permitted wavelengths, we obtain.

$$\lambda_n = \frac{2L}{n} \text{ (where } n \text{ is an integer)} \tag{3.2}$$

The kinetic energy of the particle can be expressed as

$$KE = \frac{mv^2}{2} = \frac{(mv)^2}{2m} = \frac{h^2}{2m\lambda^2} \text{ (Substituting } mv \text{ from 3.1)} \tag{3.3}$$

As the particle has no potential energy, its total energy, upon combining Eqs. 3.2 and 3.3, can be expressed as

$$E_n = \frac{(nh)^2}{8mL^2} \quad n = 1, 2, 3, 4, \tag{3.4}$$

Obviously, therefore, the restriction on its wavelength also restricts the energy that the particle can possess. Each permitted energy is identified as an energy level, and the corresponding n is called its quantum number. An important inference that can be drawn from this equation is that Planck's constant being so small ($h = 6.64 \times 10^{-34}$ joule-sec) energy quantization becomes relevant only when m and L are also sufficiently small, a signature of the microscopic world comprising of electrons, atoms, molecules, and the likes.

To drive this point home, let us consider the typical case of a football of mass 0.1 kg moving between the two walls of a room spaced by 10 m. Upon plugging these values into Eq. 3.4, we obtain the minimum energy of the ball, which corresponds to $n = 1$, to be $\sim5 \times 10^{-69}$ J. A 0.1 kg football possessing this KE will move with a velocity of $\sim3 \times 10^{-33}$ m/s, which is vanishingly small and is practically indistinguishable from the one that is at rest! If this ball has to travel at a modest speed of 1 m/s, the corresponding energy level will have a quantum number on the order of 10^{33}. In the limit of such enormously high quantum numbers, the successive energy levels are positioned relatively so close, resulting practically in an energy continuum.[1] Thus, in the macroscopic world quantization of energy becomes inconsequential and Newtonian mechanics prevails.

---

[1] A continuous sequence in which the adjacent energy levels are not perceptibly different from each other

Fig. 3.2 The emission of
the fluorescent light after
dispersion through a prism
is scanned by a detector. The
intensity recorded by the
detector as a function of the
wavelength of emission is
also shown here for a narrow
spectral range. In reality, the
spectrum will obviously be
much richer containing
many more emission lines

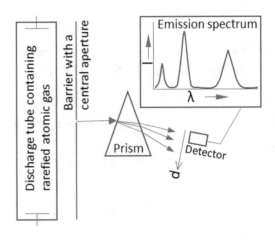

3.3   Spectral Emission from an Atomic Source

The fluorescent light sources are now in extensive use for illuminating our household
and eye-catching displays at night throughout the world. The emission from such a
source holds a wealth of science, the gradual decoding of which allowed physics to
turn the sharp corner, stemming largely from the discovery of the photoelectric effect
and a host of other great advances that soon followed. It is imperative at this point to
delve a little into the emission of this source. To this end, we perform a simple
experiment schematically described in Fig. 3.2. In fluorescent light (elaborated
in Sect. 4.2, Chap. 4), an electric discharge is impressed upon a rarefied atomic
gas contained in a glass tube. This causes the tube to emit light spontaneously all
around it. By placing an opaque barrier with a central hole to one side of the tube, we
select a narrow beam of light and allow it to disperse through an appropriately
located prism. The intensity of this spatially dispersed light is then monitored by
placing a detector on the other side of the prism and scanning it across the length d of
the spatial spread. Based on the knowledge that we acquired on prismatic dispersion
in the previous chapter, it is apparent that the spatial splitting of the beam is basically
a spread of all the wavelengths it is made up of in space. It is therefore equivalent to
saying that the distance d through which the detector is scanned can be directly
translated to the wavelength λ. This exercise thus presents a record of the intensity of
light emitted by the source as a function of its wavelength, known as the "emission
spectrum" in the common parlance.

An oversimplified emission spectrum over a narrow range of wavelengths is also
shown in Fig. 3.2 for clarity as well as palatability. A closer look at this spectrum
readily points to the fact that the fluorescent source, under study here, basically emits
light of three primary wavelengths over the displayed spectral range. Distinct also is
the fact that the emission never occurs sharply on any of these wavelengths but rather
spread into multiple wavelengths distributed around them. However, the feature that
makes the emission of light by an atomic source most remarkable is the fact that it

Fig. 3.3 Line center (λ_0) and the full width at half maximum (FWHM, $\Delta\lambda$) of the emission

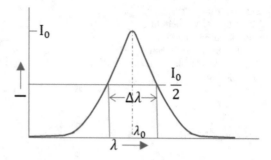

does not emit light of continuous wavelength, and, on the contrary, an element of discreteness is strikingly intrinsic to its emission spectrum. The atomic spectrum was first observed in 1853 by a Swedish physicist, Anders Angstrom (1814–1874), regarded as the founder of spectroscopy, for the hydrogen atom that would later play a pivotal role in providing the first insight into the structure of an atom.

The wavelength at which the intensity attains a peak I_0 is called the line center λ_0, and the spread around it is called the emission width $\Delta\lambda$ and is usually defined as the width of the emission at half of its peak value, abbreviated as FWHM. The same has been pictorially illustrated in Fig. 3.3 for one of the emission wavelengths of the atomic source. The emission width is a direct measure of the monochromaticity or purity of the color of the corresponding emission. It is thus at once obvious that of the three different wavelengths of Fig. 3.2, the emission at the longest wavelength is least monochromatic and that at the smallest wavelength is the most. While the key features of the emissions remain invariant with different atomic sources, the number of emitted wavelengths and their corresponding emission widths vary from one atomic gas to the other. Gaining insight into the seemingly intricate physics underscoring the atomic spectra, which swayed some of the finest scientific minds over the years, is considered a major milestone in the evolution of science.

3.4 Bohr's Atom and Beyond: A Unique Handshake Between Matter and Radiation, the Two Faces of Nature

The classical physics pictures an electrically neutral atom as a tiny but enormously massive positively charged nucleus surrounded by electrons, at a great distance away, in numbers to exactly match the central positive charge. This notion, which emerged primarily from Rutherford's famed α-ray scattering experiment, however, suffers from an inherent limitation of rendering the atom unstable. An electron cannot be stationary as it would then be pulled by the positively charged nucleus into it so rapidly that the stable atom would collapse in just a flitting second. The atomic stability demands that the electron must orbit around the nucleus, akin to the

Fig. 3.4 Schematic
illustration of an encircling
electron, which, upon
radiating energy, will spiral
inward to impact the nucleus
in a split second

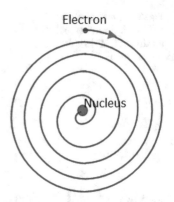

Electron

Nucleus

rotation of planets around the Sun, and the attractive Coulomb force between the
nucleus and electron provides the required centripetal force.[2] The mere rotation,
however, cannot provide stability to the atom. According to classical electrodynamics, a rotating electron would lose its energy by emitting electromagnetic waves
[19]. As a result of this energy loss, the orbiting radius of the electron reduces, and,
in turn, it spirals inward colliding with the nucleus almost instantaneously (Fig. 3.4).
Moreover, during the process of this atomic collapse, the electron must emit
electromagnetic waves of continuously ascending frequency as its speed of rotation
progressively increases. These are in stark contrast to the fact that atoms are indeed
stable and, as we have seen, the atomic emission is discrete in frequency or
wavelength and not continuous. In an attempt to resolve this impasse, Niels Bohr,
having realized the inaptness of classical physics to describe the behavior of
microscopic particles such as atoms, put a bold step forward in 1913 [4]. His
courageous assertion that the energy levels of an atomic electron are quantized,
and it can reside only in these levels, termed stationary states, without any dissipation of energy, confers stability to the atom. It is of interest to note here that the
quantization of energy levels is a direct consequence of the wave nature of matter, a
fact conceptualized by de Broglie, a decade later, in 1924. Example of waves
impressed in a wire loop as illustrated in the traces of Fig. 3.5 will help you capture
the underlying physics here. A wave, unless it joins itself as it travels around the loop
(Fig. 3.5a), cannot be sustained as the destructive interference, the occurrence of
which is inevitable here, will cause the vibration to die out. Consequently, the given
wire loop is forbidden for the wave of this wavelength. It becomes obvious that only
those waves can survive in the wire loop whose wavelength times an integer fits
exactly into its circumference (Fig. 3.5b and c). Drawing an analogy, it becomes

---

[2]Centripetal force acts on a body that performs a curved motion and is directed toward the center of
rotation. In case of planetary motion, the gravitational attraction between the Sun and the planet
provides the centripetal force. If you tie an object to one end of a rope and rotate it by holding the
other end, the tension developed in the taut rope supplies the centripetal force necessary to keep it
moving. Centripetal force in case of a particle of mass m rotating at a velocity v in a circular path of
radius r is given by $\frac{mv^2}{r}$.

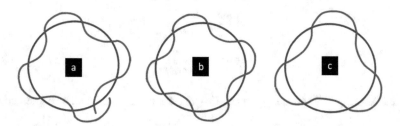

Fig. 3.5 Illustration of waves of different wavelengths vibrating in a circular loop of wire. (**a**) The vibration will die out due to destructive interference if the wave does not join on itself in the loop. (**b**) and (**c**) Only the waves whose wavelength fits an integer number of times within the loop will survive. Clearly, the values of the integer are 4 and 3 for (**b**) and (**c**), respectively

Fig. 3.6 Schematic of a hydrogen atom where the electron is revolving around the proton with a velocity v in the nth stable orbit of radius r_n

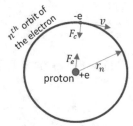

readily apparent that an electron can also circle around the nucleus without any dissipation of energy only in those orbits that contain an integral number of its de Broglie wavelengths. Little surprise then that this notion of a wavy particle allowed tying up all the loose ends and physics quickly fell into place. The wave nature of a particle nevertheless gives the atom both its stability and size. Taking a cue from the analogy of the waves in a wire loop, the radius of the possible orbits of the electron could now be readily linked to its de Broglie wavelength λ mathematically as.

$$2\pi r_n = n\lambda \tag{3.5}$$

where n is an integer, called the quantum number, and r_n is the radius of the nth stable orbit of the electron (Fig. 3.6). To simplify the analysis, we consider the case of a hydrogen atom, wherein a lone electron revolves around a proton. Upon substituting λ from Eq. (3.1), we obtain.

$$r_n = \frac{nh}{2\pi m v} \tag{3.6}$$

where m and v are the mass and velocity of the electron, respectively. The centripetal force F_c required to hold the electron in its orbit is provided by the electrical attraction F_e between the proton and the electron each containing a charge of magnitude e. It therefore readily follows that

$$\frac{mv^2}{r_n} = \frac{e^2}{4\pi\epsilon_0 r_n^2}$$

$$\text{i.e. } v = \frac{e}{\sqrt{4\pi\epsilon_0 m r_n}} \tag{3.7}$$

Eliminating "v" between (3.6) and (3.7), r_n, the orbital radii, can be expressed as.

$$r_n = \frac{n^2 h^2 \epsilon_0}{\pi m e^2} \tag{3.8}$$

The corresponding energy "E_n" of the electron will be the sum total of its kinetic and potential energy,[3] i.e.,

$$E_n = KE + PE = \frac{mv^2}{2} - \frac{e^2}{4\pi\epsilon_0 r_n}$$

Substituting "v" and "r_n," respectively, from Eqs. (3.7) and (3.8), we find that.

$$E_n = -\frac{e^2}{8\pi\epsilon_0 r_n} = -\frac{me^4}{8\pi\epsilon_0^2 n^2 h^2} =$$

$$-\frac{E_1}{n^2} \left(\text{where } E_1 = \frac{me^4}{8\pi\epsilon_0^2 h^2} \text{ is the energy in the first orbit} \right) \tag{3.9}$$

Obviously, therefore, each stable orbit of the electron has a discrete energy associated with it. This fact has been illustrated in Fig. 3.7 for the first three stable orbits of the electron. The negative energy of the electron implies that the electron is bound by Coulomb attraction and work needs to be expended to make it free. For instance, the electron revolving in the first orbit has a negative energy of magnitude E_1, and, therefore, to make it free exactly E_1 energy must be supplied from outside. A free electron is no longer bound by the attraction of the proton and thus will possess zero energy that, as evident from Eq. 3.9, corresponds to a quantum number of infinity. All the remaining energy levels lying between $n = 3$ and $n = \infty$ would pack to the capacity this slender energy space. This fact has also been highlighted in Fig. 3.7 by providing an enlarged view of this otherwise narrow energy space. As seen, with increasing n, the energy levels approach each other so closely that they eventually merge practically into a continuum. This is understandable; as the electron tends to be free, its de Broglie wavelength is no longer constrained with the so-called boundary conditions, and quantization of energy becomes practically irrelevant.

---

[3] Potential energy of the electron here is negative as it is being electrically attracted by the proton and equals the work required to be imparted in order to move the electron through a distance between r_n to infinity.

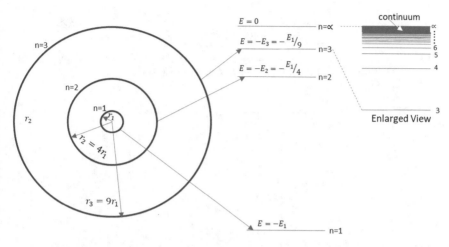

Fig. 3.7 First three electronic orbits in the hydrogen atom drawn to the scale of proportionality to n^2. The corresponding energy levels are shown alongside and is drawn to the scale of inverse proportionality to n^2. The zero-energy state of the electron obviously corresponds to a quantum number of infinity. An enlarged view here shows the stuffing of the energy space between $n = 3$ to ∞ with increasingly closely spaced energy levels leading eventually to a continuum

The very conceptualization of quantization of energy by Bohr, coming at a time when de Broglie was yet to enlighten the scientific world on the wavy behavior of the matter, was indeed a masterstroke and left an indelible mark in the evolution of knowledge. The perception of the energy quantization laid the groundwork for the interpretation of the atomic spectra that had eluded the scientific community for well over half a century. Bohr's exposition that the atomic spectra originate from a unique handshake between the wavy particle and corpuscular wave, the two most guarded secrets of nature until the beginning of the twentieth century, startled the scientific community. Many expressed their disbelief, and Einstein was no exception. A few such as Otto Stern (1888–1969) and Max von Laue (1879–1960), both German by birth and Nobel Laureates in Physics, offered to quit physics if Bohr were correct. That they had to retract the offer later is of course another story. Bohr's interpretation of atomic spectra in conjunction with de Broglie's formulation of matter waves provided the initial impetus for quantum mechanics to navigate the challenges of the microscopic world.

As we know now, the wavy electron in the hydrogen atom, under the influence of the proton's electric field, is able to reside without dissipation only in certain orbits, each labeled by a different integer called the orbital quantum number. As the electron is bound here, its total energy is always negative. An electron in the innermost orbit, identified as the orbit with quantum number $n = 1$, possesses the least energy and can be determined from Eq. (3.9) to be -13.6 eV.[4] The energy of the electron rapidly

---

[4] eV is normally the unit used to express the energy of atomic particles and 1 eV is equivalent to 1.6×10^{-19} J.

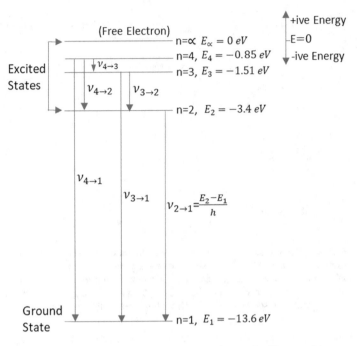

Fig. 3.8 All the stable atomic orbits of the electrons are characterized by negative energy. A few low-lying energy levels of a hydrogen atom are illustrated here along with their negative values of energy. Origin of spectral lines as the excited electron makes a transition to an energy level with lower energy is also schematically depicted

increases with increasing n and attains a value of zero as n tends to infinity. A few low-lying energy levels of a hydrogen atom and the corresponding energies are shown in Fig. 3.8. The electron under normal circumstances occupies the level with the lowest energy called the ground state. When energized by an extraneous source, the electron is excited and climbs up the energy ladder. Upon receiving an energy of 13.6 eV or more, the ground state electron will become free and no longer remain bound to the proton to form the hydrogen atom. This is equivalent to saying that the hydrogen atom has been ionized as the electron is stripped off. However, the plot thickens if the ground state electron acquires energy not sufficient to tear off the proton's attraction and escape, but just enough for it to jump into another low-lying energy level instead. For instance, what happens if the electron has climbed to the second orbit from the first: a gripping point that seemingly intrigued Neil Bohr's mind. His surmise of the electron spontaneously dropping down to the ground state releasing its energy of excitation as a photon of light is regarded as a brilliant piece of work as it finally cracked the physics behind the discreteness of atomic spectra. Digging a little deeper, we can readily conclude that this interpretation points to an exceptional hand clasp of particle and wave; it's like the wavy atom has created energy steps, akin to a ladder, for the photonic wave to climb down to the ground. If the electron makes a quantum leap from the second orbit ($n=2$, and energy $= -E_2$) to the first (n=1, and energy = $-E_1$), the energy of the photon emerging from this transition will equal the difference of energy between these two levels. This can be mathematically expressed as

$$hv = -E_2 - (-E_1)$$

$$v = \frac{1}{h}(E_1 - E_2) \tag{3.10}$$

where v and h are the frequency of the emitted photon and the Planck's constant, respectively. If the electron, on the other hand, is excited to the next higher orbit of quantum number 3, it can return to the ground state by following two different routes, either making a direct transition from 3 to 1 or via level 2. This would thus result in the emission of three different photons. Clearly, the higher the level of excitation, the more numerous the pathways for the electron to return to its ground state. Every transition that the electron makes over its ground state-bound journey gives rise to the emission of a photon of different wavelength. The discreteness of the wavelength of light is thus intrinsic to the atomic spectra. Bohr's theory, however, neither sheds any light on how long the electron can reside in an excited state before making the spontaneous downward leap, nor does it say if some transitions are more probable than others or on the forbiddenness of any transition. The answers to these questions not only are central to understanding the other aspects of the atomic spectra, such as its varying richness from atom to atom, or as to why the width of one transition differs from another, but also have a strong bearing on the operation of a laser. These will be addressed at another place in the book.

3.5 Boltzmann Distribution

We now know that quantization of energy is a rule of nature that the inhabitants of the microscopic world very compliantly obey. Not that it doesn't exist in the macroscopic world, but we are unable to realize this as its effect becomes vanishingly small for the conventional objects. In addition to providing physical insight into the origin of quantization, the preceding sections also offered a quantitative formulation of this in the case of a hydrogen atom, the simplest of all the atoms. Although the theoretical approach to be followed in this context for atoms with higher atomic numbers will be qualitatively similar, the presence of multiple electrons and their associated de Broglie wavelengths will make the analysis understandably more complex. Notwithstanding this, the fact remains that no atom, small or large, can disobey the rule of energy discretization. The same will also apply to atoms from which one or more electrons have been stripped off. A molecule is an amalgamation of multiple atoms bonded together and will, nevertheless, have its own quantized electronic energy levels similar to that of an atom. As will be shown in a latter section of the book dealing with molecular lasers, a molecule, unlike an

Fig. 3.9 Distribution of
total population "N"
between two energy levels
satisfying Boltzmann's law

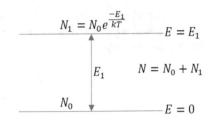

atom, can also vibrate and rotate at the same time. Similar to its electronic energy levels, the molecule's vibrational and rotational levels are also quantized. An ensemble of atomic or molecular systems will comprise countless number of atoms or molecules. The obvious question that arises here is how these numerous species will be distributed among the discrete electronic energy levels in the case of atoms and the electronic, vibrational, and rotational energy levels in the case of molecules? To simplify the matter, let us stick at this point only to the case of distribution of species, atomic or molecular, among one set of discrete energy levels. (The specific cases of distribution of atoms and molecules into their respective energy levels will be addressed in a latter chapter dealing with atomic and molecular lasers.) Let us consider a situation wherein a total of N species are to be distributed within two energy levels as illustrated in Fig. 3.9. One of the energy levels is the ground level that for convenience has been scaled to a zero-energy state and the other is located at an energy of E_1 above it. It is a well-known fact that nature always acts in a manner so as to establish a situation with the lowest possible energy. In this example, the minimum energy condition will obviously be the one where all the N species settle into the ground energy state. There is a catch though! For any nonzero temperature, the system will contain a finite amount of thermal energy that essentially manifests by raising a fraction of the species to the levels with higher energy. In 1877, Ludwig Boltzmann[5] (1844–1906), an Austrian physicist, succeeded in theoretically predicting the relation of the equilibrium population N_1 of any energy level to the ground level population N_0 as a function of the temperature, as

$$N_1 = N_0 e^{\frac{-E_1}{kT}} \text{ where } k \text{ is the Boltzmann's constant and } T \text{ is the temperature}$$

In reality, however, there would be an infinite number of energy levels, and consequently, the above formula can be generalized as

---

[5] Boltzmann chose to die by his own hand on September 5, 1906, when he was holidaying with his wife and daughter at a place close to the Italian town of Trieste. His suicide is blamed to a severe depression, he was suffering from, that stemmed basically from the strong opposition to his work on theorizing the population distribution. The irony is only a few weeks following this tragedy, his theory was experimentally verified. The suicide most certainly also drew a curtain on his getting a Nobel Prize that, incidentally, is not awarded posthumously.

$$N_j = N_0 e^{\frac{-E_j}{kT}} \tag{3.11}$$

where N_j is the population of the jth energy level whose energy relative to the ground state is E_j.

In equilibrium, the population will be distributed among all these levels satisfying this equation termed Boltzmann's distribution law. It may be possible to disturb this state of thermal equilibrium by some extraneous means, but the condition of nonequilibrium will be short-lived as the system rapidly readjusts itself to return to its original equilibrium state. It becomes readily apparent from Eq. 3.11 that for any finite temperature T, population of an energy level exponentially reduces with the increase in its energy relative to the ground state. It is equivalent to saying that it is impossible to have an energy level with population exceeding that of any level beneath it. This is a valid statement for any highest conceivable temperature. Even in the limit of temperature tending to infinity, its population can at most match that of the level lying beneath and can never exceed it. In fact, in that limit, the population of all the levels will match that of the ground state. On the other hand, if the temperature starts falling, the low-lying levels will begin becoming increasingly populated at the expense of those in the upper levels. If the temperature goes all the way down to absolute zero, the entire population of the system will be realized in the ground state. The bearing of these facts in the realization of population inversion and, in turn, operation of a laser will be addressed in the next chapter.

Chapter 4
Lasers: At a Glance

4.1 Introduction

The discrete nature of energy levels and the transition of atoms from excited to lower levels are, as we know now, the origin of quantum light sources that we so commonly make use of for illumination and eye-catching displays at night. Quantum sources of light are also employed in the laboratory for many light-based experiments. Instances are aplenty when the outcome of such experiments has compelled the enigmatic nature to unwillingly unravel some of its deepest secrets. For instance, an explanation of the photoelectric effect, discovered in the beginning of the last century when ultraviolet light was made to shine on a metal surface, as perceived by Albert Einstein, cemented the photonic concept of light. Incidentally, as the majority of readers are perhaps aware of, Einstein was awarded the Nobel Prize in Physics for unmasking the physics underneath the photoelectric effect and not for his earthshaking work on relativity. It is also well known that Chandrashekhar V. Raman (1888–1970) made use of a mercury vapor lamp in his experiments when he showed that upon scattering from an appropriate medium, the photon energy can either increase or decrease, a discovery that fetched him a Nobel Prize in physics. We shall take a closer look at his ingenious experiment in a future chapter (V-II). Just as it occurs in such light sources, in a laser too, a transition of an excited species to a lower energy level is the basis of light emission. However, what makes the laser so distinctly different from this other kind of sources? To unravel this, we need to first understand the working of a common quantum source of light. The purpose of this chapter is to further this understanding toward imparting a rudimentary impression of the laser into the mind of the readers without compromising the scientific accuracy. The anecdote picked from the history enveloping the invention of laser by Maiman[1] provides a seamless ending to this

---

[1] Interested readers may refer in this context to the book *The Laser Odyssey* by Theodore Maiman, creator of the world's first laser.

© The Author(s), under exclusive license to Springer Nature Switzerland AG 2023   73
D. J. Biswas, *A Beginner's Guide to Lasers and Their Applications, Part 1*,
Undergraduate Lecture Notes in Physics,
https://doi.org/10.1007/978-3-031-24330-1_4

stand-alone chapter. Subsequent chapters (5 to 12) of this part of the book are devoted to unveiling the laser more comprehensively while preserving the intuitive approach of this endeavor.

4.2   Working of a Conventional Fluorescent Light

A typical fluorescent tube light that illuminates our surroundings in the dark is schematically illustrated in Fig. 4.1. A glass discharge tube here contains some neutral atomic gaseous species that we call A at a low pressure, typically, a few

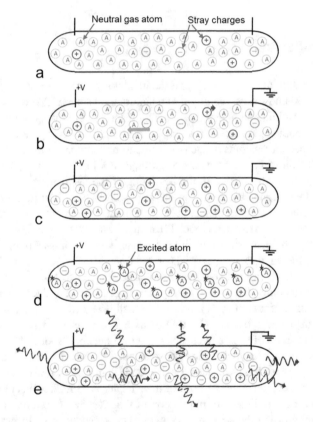

Fig. 4.1 Schematic representation of the mechanism of operation of a conventional fluorescent light source (**a**) The glass discharge tube containing the neutral gas atoms and a few stray electrons and ions (**b**) On application of voltage V, the charges are accelerated. Electron, being much lighter, is accelerated faster acquiring higher kinetic energy (**c**) The accelerated electrons with appropriate KE on collisions with neutral atoms result in their ionization creating an avalanche of electrons (**d**) Some of the collisions of the electrons with neutral atoms may be inelastic wherein the atom gets electronically excited (**e**) The excited atoms return to the ground state releasing the excitation energy in the form of photons emitted randomly

torr or so. Although electrically neutral, the presence of at least a few stray charges in the form of electrons and cations in the gas, however small, is inevitable[2] (Fig. 4.1a). Upon application of an appropriate voltage V across the two electrodes placed at the two ends of this tube and separated by a distance d, both the electrons and the positive charges experience a force that results in their acceleration, the electrons toward the anode and the positive ions toward the cathode (Fig. 4.1b). The electrons being much lighter experience significantly higher acceleration given by eE/m_e, where $E = V/d$ is the applied electric field, m_e is the mass of the electron, and e is the charge on it. As the electron moves, it gains velocity and thus acquires kinetic energy. If the electron, after acquiring sufficiently high kinetic energy, collides with a neutral atom A, it may then succeed in its ionization, leading to the creation of an additional electron that, in turn, participates in the process of further ionization (Fig. 4.1c). This avalanche effect results in a copious supply of electrons leading to the flow of current. The ions, being massive compared to the electrons, have negligible contributions to the flow of current. Some of the collisions of the electrons with the neutral atoms can also be inelastic. This means that a part of the kinetic energy of the electron is transferred to the atom as its internal energy, and the atom, in turn, is electronically excited (Fig. 4.1d). The excited atom decays to the lower energy state after a finite time releasing this energy of excitation as a quantum of light, the so-called photon. Since these are random emissions, the resulting photons would travel in all possible directions around the discharge tube, i.e., over the entire 4π solid angle around it (Fig. 4.1e). The light from this device would, therefore, spread everywhere and hence is most suitable for the purpose of viewing. In practice, conventional tube lights contain mercury vapor, and the walls are coated with fluorescent paint to absorb the photons emitted in the discharge and reemit in the visible region.

4.2.1 Is It Possible to Force an Excited Atom to Emit Photon in a Particular Direction?

An obvious question that arises here is if it would be possible to force all the electronically excited atoms in "Fig. 1d" to emit in the same direction, say along the axis of the discharge tube. If it indeed happens, it is not difficult to imagine that the beam of light that now emerges would be enormously more intense (Fig. 4.2). The answer to this question is an emphatic *yes*, thanks to the deep insight of Albert Einstein, who envisaged the possibility of stimulated emission way back in 1917. However, it took more than four decades for the experimental exploitation of stimulated emission to manufacture such a unique beam of light. The name Einstein is synonymous with $E = mc^2$ or the *theory of relativity* and may be to some extent

---

[2]Cosmic radiation that reaches the Earth surface from the outer space is responsible for the presence of charged particles in our atmosphere.

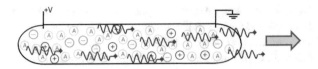

Fig. 4.2 If all the excited atoms of Fig. 1d were to emit in a particular direction, a powerful beam of light would then emerge. The device, however, ceases to remain a viewing source anymore

with the explanation of the photoelectric effect that an informed reader might associate with his Nobel Prize winning piece of work. However, it is much less well known that the innovative concept of stimulated emission, which has given birth to a device that now has practically pervaded every walk of our life, is also his brainchild. Before venturing into the realization of stimulated emission in an experimental system, it would be useful to briefly describe the spontaneous and stimulated emission processes.

4.3   Spontaneous and Stimulated Emissions

Let us consider an atom in an excited electronic state. This energy of excitation can be imparted by shining it with a photon of just the right amount of energy or through an inelastic collision with an electron possessing an appropriate amount of kinetic energy, as is the case of a fluorescent light source. After spending a finite amount of time in the excited state, the atom eventually comes down to the ground state, releasing this energy of excitation as a photon. The photon can be emitted anywhere within the entire 4π solid angle around this excited atom. Such an emission, which makes it possible for a viewing source to spread its light everywhere, is called spontaneous or random emission, and the same is qualitatively depicted in Fig. 4.3. Let us once again consider an atom in the excited state that is in the verge of spontaneously emitting a photon of frequency ν. However, in the event of its interacting with a photon (I) of the same frequency ν, while it is still in the state of excitation, it would instantaneously come down to the ground state, emitting a second photon (II). Furthermore, this second photon is identical to the first photon (I), which interacts with the excited atom, called the stimulating photon, in all respects. Most importantly, it travels in the same direction as the first photon. Such an emission is called forced, induced, or stimulated emission and is schematically depicted in Fig. 4.4. It is obvious from this discussion that the seed for stimulated emission is spontaneous emission itself.

Returning to the example of tube light, the photons due to spontaneous emission will be emitted by the excited atoms in all possible directions. However, the photons that happen to travel along the axis of the tube light will have the opportunity to stimulate the maximum number of excited atoms to emit identical photons along the same direction. This is not the end of the story. All these photons emitted due to stimulated emission, traveling along the axis by default, will cause further

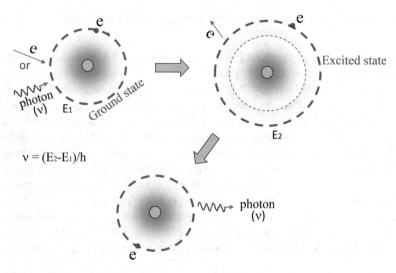

Fig. 4.3 Schematic representation of spontaneous emission. An atom having absorbed a photon of appropriate frequency or inelastically scattered by an electron goes into the excited state. After a finite residence time, the excited atom returns to the ground state releasing the excitation as a photon

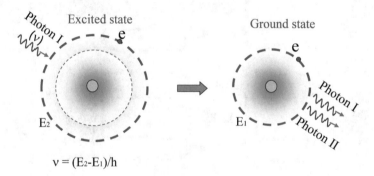

$$\nu = (E_2 - E_1)/h$$

Fig. 4.4 Schematic representation of stimulated emission. The atom may reach the excited state in the same way as depicted in Fig. 4.3. When a photon of appropriate energy interacts with this excited atom, it instantly comes down to the ground state emitting a photon that is identical in all respect to the photon that caused this de-excitation

stimulation of excited atoms on their path, leading to a multifold rise in their number. Thus, ordinary tube light should also be able to give an intense beam of light as a result of amplification of those few spontaneously emitted photons along the tube axis. Surely this is not true; otherwise, the laser would have come into existence more than a century ago along with large discharge length tube lights. Where then lies the fallacy??

4.4 Population Inversion and Amplification of Light

Let us consider the simple case of an ensemble of atoms that are capable of residing either in the lower or excited energy states (Fig. 4.5). Under normal conditions, the majority of the atoms occupy the lower energy state, as depicted in Fig. 4.5a, following the Boltzmann distribution, we studied in the preceding chapter. We shine this medium by a beam of light comprising photons of energy $h\nu$ that exactly matches the difference in energy between these two levels, called resonant photons. As these photons travel through the medium, they may interact with the population both in the lower or excited energy levels. If a photon interacts with an atom in the lower state, it is absorbed, and the atom in turn is excited. On the other hand, if it interacts with an excited atom, it stimulates it to return to the lower state emitting in the process a second photon, identical to the first, the so-called stimulated emission. However, as the number of atoms in the lower state is greater than that in the excited state, the absorption of photons will exceed the number of photons emitted due to

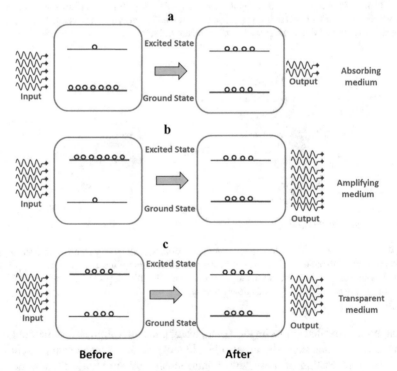

Fig. 4.5 Schematic illustration of the passage of a beam of light through a resonant medium (**a**) Normal case where majority of population resides in the ground state. Understandably, the medium behaves here like an absorber (**b**) A special case where population in the excited level exceeds the ground state population. The medium here is conducive for amplification of light by stimulated emission (**c**) Another special case when the population of the excited and ground states are equal. The medium here behaves like a transparent window

stimulated emission. Thus, the incident beam, as it traverses through the medium, is gradually diminished in the number of photons, and an attenuated beam finally emerges (Fig. 4.5a). The medium therefore behaves as an absorber here. At this point, we consider the reverse situation wherein the population in the excited level exceeds that in the lower level (Fig. 4.5b). The traversing beam in this case interacts more with the atoms in the excited state than with those in the lower state, causing the stimulated emission to outnumber the absorption. The net effect would therefore be amplification of the input beam (Fig. 4.5b). Thus, we see that amplification of light by stimulated emission of radiation would be possible if and only if the population of the excited energy level exceeds that of the lower level, a condition that is normally termed population inversion. However, such a situation violates the Boltzmann equilibrium and thus cannot come about naturally. In reality, therefore, the population in the excited level is always lower than that in the lower level unless a mechanism to invert the population,[3] a trick not so easy to realize, is in place. In an ordinary fluorescent tube light, although the electric discharge succeeds in transferring a fraction of the ground state population to the excited state, the condition of population inversion is not met. The amplification of light by stimulated emission is thus ruled out, resulting in the predominance of spontaneous emission spreading the light in all directions. It would be pertinent here to consider a third situation, as represented in Fig. 4.5c, where the population of the two levels is equal. The chance of the photons of the incident beam interacting with the population in the two levels being the same, it is understandable that the incident light beam now traverses through the medium unaffected. The medium now behaves like a transparent window. This situation, like the case of population inversion, is also a violation of the Boltzmann distribution, and the ways of experimentally realizing a population inverted condition will be addressed in the following chapter.

Thus, a light amplifier can be pictured much the same way as Fig. 5b, where there is a medium in which the population is inverted between two energy levels and an input beam of identical resonant photons of energy matching exactly with the energy difference of these two levels passes through it. The medium, by virtue of stimulated emission, adds more identical photons to this incident group of photons and, in turn, gives out an amplified beam of light. The higher the population inversion is, the greater the amplification. Thus, we can say that the gain of this optical amplifier increases with increasing population inversion. Laser, we now know, is the acronym for light amplification by stimulated emission of radiation. Therefore, it is also an amplifier of light but surely does not require an input source for its operation. If it did, then the laser would have lost all its charm. How does it work then? It amplifies light and yet it does not need an input for amplification! To understand this, we must know what a cavity is.

---

[3] This can be achieved if somehow the population is selectively transferred from the source to the excited level.

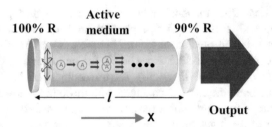

Fig. 4.6 Schematic representation of the multiplication of photons due to stimulated emission, initiated by the emission from a single excited atom inside a simple cavity. The black dots represent continuation of the same process

4.5   A Simple Cavity

A simple cavity is formed when two high reflectivity mirrors are placed parallel to each other at a given separation. Here, we consider one of the mirrors to have 100% reflectivity, while the other is 90% reflective, meaning that it has a transmission of 10%. The population inverted medium, also called an active medium, is placed inside this cavity (Fig. 4.6). The excited atoms spontaneously decay to the ground state, emitting photons in all directions, as shown in the figure. In this figure, we have tagged one such atom A, at the extreme left end, that happens to have emitted a photon along the axis of the cavity in the positive X direction. As it travels, the photon encounters another excited atom A and stimulates it to emit an identical photon. Thus, we have two identical photons traveling along the cavity axis that are ready to cause two more stimulations, making the number of identical photons moving along the axis four. This process continues taking the number of identical photons to astronomical values.[4] This happens only because the number of excited atoms exceeds their number in the ground state in an active medium. The point that needs to be emphasized here is that while we have considered only a single excited atom at one end of the medium, this holds true for innumerable excited atoms lying anywhere in the active medium and emitting axially. It is also not difficult to visualize that the same story holds true for every excited atom emitting photon in the negative X direction. The photons traveling in the –X direction are reflected off the 100% R mirror; travel through the inverted medium again, causing further multiplication of their number; and reach the partially reflecting (90% R) mirror where the majority of them are reflected back. The back-and-forth oscillation of these photons and their repeated passage through the inverted medium culminate in the generation of a huge photon flux inside the cavity, a part of which continues to be emitted through the partially transmitting mirror as the laser output. The spontaneous emission in any direction other than the axis (as depicted in Fig. 4.6) cannot grow in

---

[4]The increase in the number of photons is exponential and is proportional to e^{gl} where gain g is directly proportional to population inversion and l is the length of the active medium.

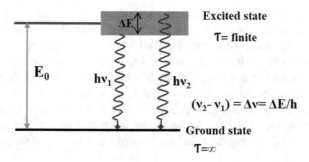

Fig. 4.7 Finite lifetime associated with an excited energy state leads to its broadening in energy in accordance with Heisenberg's uncertainty principle. The residence time in the ground state being infinite has no such broadening. The energy broadening leads to a corresponding spread in the transition frequencies

this manner due to the absence of a cavity.[5] It is easy to understand that to obtain a continuous beam of light from this device, the mechanism for creating population inversion, also called "pumping," must also be operational in a continuous manner (readers are referred to Chap. 7 for more comprehensive knowledge on this). Obviously, therefore, a laser does not require an input source of light for its operation. The cavity enables the spontaneously emitted photons to perform the role of input seed photons for amplification through stimulated emission.

It is well known that an atom has a finite residence time in the excited state before it decays spontaneously to the ground state (readers may also refer to Chap. 8 in this context). Invoking Heisenberg's uncertainty principle once again but now with respect to canonically conjugate variables,[6] time and energy, it is straightforward to show that the finite residence time leads to a corresponding spread in the values of energy that the excited atom can possess, aptly termed the broadening of the energy levels. This situation has been illustrated in Fig. 4.7. This spread in energy of the excited state will result in a corresponding spread in the frequency of the spontaneously emitted photons. Since the stimulated emission is seeded by these spontaneous photons, one would also expect a similar frequency spread in the emission of the laser. If true, then the color of the laser light would be far from pure. The cavity does not allow this to happen, as it lets to and fro oscillations only for those wavelengths, half of which times an integer fits exactly inside it. These discrete wavelengths or corresponding frequencies are called cavity modes and have been elaborated in greater detail in Chap. 6. The cavity, by virtue of its discretionary power, would allow only those photons to grow through stimulated emission whose frequency matches that of the cavity modes and thus helps the laser to preserve the purity of its color. Unquestionably, therefore, the integration of a cavity with stimulated emission

---

[5]The amplification in certain cases can be extremely high in a single pass itself. This is called amplified spontaneous emission.

[6]If the product of dimension of any two variables equals the dimension of the Planck's constant h, then they are called canonically conjugate variables in the present context.

not only allows multi-pass amplification but also makes its emission both directional and monochromatic. Thus, we have now zeroed in on the two most important characteristics of lasers, viz., monochromaticity and directionality, which, in turn, lead to many other attributes that a laser is associated with as elucidated in Chap. 6. However, the very first laser that Theodore Maiman operated in 1960 did not make use of a conventional cavity.[7] One would then begin to wonder as to how amplification of light by stimulated emission manifested in his device. It would be interesting to digress a little here to know the story behind the first experimental demonstration of light amplification by stimulated emission of radiation.

4.6 Maiman's Experiment: Invention of Laser

American physicist Theodore Maiman operated the first laser, a ruby laser, on May 16, 1960, at the Hughes Research Laboratories in Culver City, California. As a tribute to this magnificent invention, May 16 is celebrated as International Day of Light all over the world. The paper that Maiman wrote to report this invention comprised just a few illustrations and a small text. It appears that there was an element of urgency to quickly disseminate his results. This is understandable considering the intense research activity that was being witnessed at that time aimed at achieving the first experimental demonstration of amplification of light by stimulated emission of radiation. It is also possible that Maiman, having obtained the results, was confident that they stood tall enough to prove the point that he was making. The paper in its rudimentary form was communicated to *Physical Review Letters (PRL)* for consideration for publication. The then-editor Sam Goudsmit misjudged the value of the work and found it to be unsuitable to warrant publication in a journal of the repute of *PRL*. This paper of Maiman was then subsequently published in the prestigious *British Journal Nature*, which has since been recognized as the first ever experimental demonstration of amplification of light by stimulated emission of radiation. It may be of interest to note here that even after submitting the paper to *Nature*, Maiman appealed to the editor of *PRL* to reconsider his decision and offered to withdraw the paper from *Nature* if he would agree to its publication. Goudsmit, however, refused to budge and, worst still, labeled this revolutionary piece of work as duplicate. In 2003, *Nature* brought out a unique book entitled *A Century of Nature: Twenty-One Discoveries That Changed Science and the World*, wherein the invention of lasers written by Charles H. Townes, a recipient of Nobel Prize for his overall contribution in the development of the coherent source of EM radiation, featured as the eighth article. In Townes' own word, "the Maiman's paper

---

[7] Although this fact was not explicitly mentioned by Maiman in the paper he wrote reporting the invention of the laser, the non-emergence of a beam of light points to the cavity less operation of the very first laser. This aspect was also corroborated by Nobel Laureate Charles Townes in his article "The First laser" in *A Century of Nature: Twenty-One Discoveries That Changed Science and the World* Nature, 2003, Ed: Laura Garwin and Tim Lincoln.

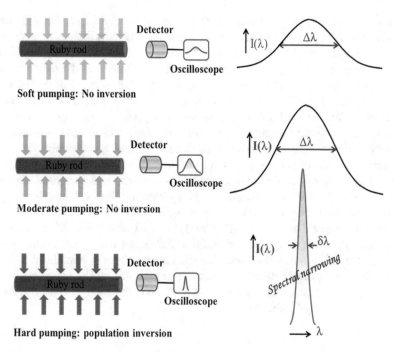

Fig. 4.8 A cylindrical ruby rod is optically pumped by a flashlamp, and the resulting fluorescence is monitored by an appropriate detector placed on one end of the ruby. At low flashlamp intensity, the spectral width of the emission matched with the known fluorescence width of ruby (top trace). At a moderate flashlamp intensity, the fluorescent intensity too increased but its spectral width was the same (middle trace). At high flashlamp intensity that caused population inversion, the significant spectral narrowing of the fluorescence provided the first ever signature of amplification of light by stimulated emission of radiation (bottom trace)

is so short and has so many powerful ramifications that I believe it might be considered the most important per word of any of the wonderful papers in Nature over the past century." This ornate observation coming from a noble laureate who himself was outdone by Maiman in the race to create the very first laser speaks volumes of the importance of this work.

Here, we present an easy-to-understand explanation of Maiman's work[8] by using an overtly simplified illustration as depicted in Fig. 4.8. A cylindrical ruby rod was optically pumped by deriving light from a flashlamp. A part of the light emitted by the flashlamp can cause resonant electronic excitation of the ruby molecule that would subsequently fluoresce light of a correspondingly longer wavelength. The

---

[8]The text and illustrations here are not a reproduction from Maiman's original research paper. To make it palatable to the readers, illustrations have been redrawn and accompanied by suitable explanation to elucidate the central theme of his work. Interested readers are referred to his original paper [T. H. Maiman, Nature 187: 493–494, August 6, 1960] to know about the invention of laser in the inventor's own words.

fluorescence has a spectral width for the same reason for which the gain has a spectral broadening, the so called Heisenberg's uncertainty principle, as elaborated earlier in this chapter. Maiman attempted to study the fluorescence as a function of its wavelength by placing an appropriate detector at one end of the ruby rod. This was done for three cases of pumping. In the first case, the intensity of the flashlamp was low, meaning that the pumping was soft. The spectral width $\Delta\lambda$ of fluorescence in this case matched the known fluorescence width (which basically is the spontaneous emission width) of a ruby crystal (Fig. 4.8, top trace). He then increased the intensity of pumping to a moderate level, thereby increasing the number density of the excited species but not good enough to create population inversion. Although the intensity of the fluorescence increased now in accordance with increased pumping, the spectral width of the fluorescence remained invariant (Fig. 4.8, middle trace). In the third case, he increased the intensity of pumping to such an extent that population inversion was created over the length of the ruby rod. In this case, Maiman observed a dramatic narrowing of the spectral width of fluorescence (Fig. 4.8, bottom trace), and he recognized this as a signature of amplification of light by stimulated emission of radiation. We provide below a qualitative explanation of the narrowing of the spectral width of fluorescence as a result of the amplification of light due to the stimulated emission of radiation.

4.7 Spectral Narrowing: A Signature of *Laser*

From the very nature of stimulated emission, it is apparent that its probability at any given frequency increases with increasing number of photons at that frequency. To understand the consequence of this, we imagine the ruby rod to be divided along its length into a number of subsections (Fig. 4.9). First, just for the sake of convenience, let us pick up such a section at the extreme left end. The distribution of photons emitted due to spontaneous emission (fluorescence) across this section with a spectral width of $\Delta\nu$ is shown immediately above it in the figure. A fraction of these photons, moving along the axis to the right, interacts with the inverted medium of the next section, resulting in their amplification. At the central frequency of the fluorescence, called the line center, where the number of photons is the highest, the amplification due to stimulated emission is also the largest. This means that the number of photons at the line center increases more rapidly than at the wings, leading to a narrowing of the spectral width. This continues as the photon distribution propagates through the sections one after the other with the narrowing effect becoming progressively more prominent as is clearly depicted in Fig. 4.9. It should be noted here that the spectral narrowing $\delta\lambda$ shown in this example is the maximum possible as the seed spontaneous photons have begun their journey from the extreme left end, traveling through the entire length of the active medium. This narrowing is thus indeed a clear signature of amplification of light by stimulated emission of radiation. In the two cases of soft and moderate pumping, in the absence of

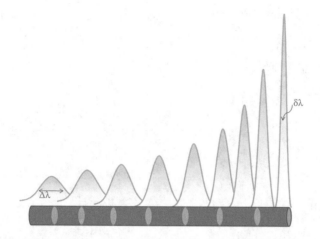

Fig. 4.9 The gradual reduction in spectral width and increase in intensity of the packet of spontaneously emitted photons over a subsection of the active medium as it traverses from extreme left to right through the length of ruby rod wherein population inversion has been created is depicted. Although a single packet of spontaneously emitted photons in the extreme left end of the ruby rod has been considered here, similar sequence of events should occur to the group of photons originating anywhere in the active medium. This decides the overall narrowing of the spectral width

population inversion, the medium acted as an absorber with no effect on the spectral width of fluorescence.

We have used very simple but revealing illustrations to bring out the central role that the population inversion and a cavity play in the operation of a laser. In the following chapters, we shall study the population inversion and cavity and their related effects in greater detail. We also have seen that the spectral narrowing effect bears a distinct signature of the amplification of light by stimulated emission of radiation. While Maiman demonstrated amplification of light by stimulated emission in an active medium without the usage of a conventional cavity, placing it inside a cavity can dramatically improve the monochromaticity of the laser. The spectral narrowing effect becomes much more pronounced in a cavity that in turn leads to much purer color from the laser, as we shall see in a future chapter.

Chapter 5
Population Inversion and Consideration of Energy Levels of a Lasing Medium

5.1 Introduction

Lasers, as we know by now, are devices that are capable of delivering, unlike any other light sources, a unique beam of light by exploiting the principle of stimulated emission. Having acquired elementary knowledge on the working of lasers by virtue of our reading through the previous chapter, we are now aware that stimulated emission alone cannot lead to the amplification of light. Else, the laser would have come into existence more than one century ago along with the development of fluorescent light sources based on a cylindrical geometry. Although it provides countless excited atoms spread along the length of the tube for their stimulation by the readily available spontaneously emitted resonant photons, yet amplification of light fails to happen. This is in spite of the fact that the avalanche of electrons produced in the electric discharge results in the creation of many excited atoms capable of giving out enough spontaneously emitted photons to break the darkness of an entire room. The crucial question here is: how many? Unless the number of excited atoms in the discharge tube outnumbers the unexcited atoms, a condition identified as population inversion in the common parlance, amplification of light is not possible. The vital role that population inversion plays in the amplification of light has been elaborated in Chap. 4 and also briefly underlined in the introductory chapter of this book. We have also studied in Chap. 3 the inevitability of quantization of energy in atoms and molecules as well as the distribution of population among these discrete levels based on Boltzmann's law. In addition to providing deeper insight into the concept of population inversion, this chapter has also been planned to enlighten readers on the inherent difficulties in its realization and some practical ways to overcome them. A significant part of the chapter is also devoted to identifying the medium most suitable for effecting and sustaining the condition of population inversion and, in turn, lasing.

© The Author(s), under exclusive license to Springer Nature Switzerland AG 2023
D. J. Biswas, *A Beginner's Guide to Lasers and Their Applications, Part 1*,
Undergraduate Lecture Notes in Physics,
https://doi.org/10.1007/978-3-031-24330-1_5

5.2   The Centrality of Population Inversion in the Context of a Laser

The crucial role played by the population inversion in the amplification of light has been touched upon in at least two of the previous chapters. The discrete nature of the energy possessed by the atoms or molecules and the richness of their interactions with light, introduced before, will be recalled here to unwrap the physics blanketing the concept of population inversion. Consider an oversimplified situation of two energy levels between which an atom (or a molecule or an ion) can make transitions upon interacting with a photon of energy matching exactly the energy difference of the two levels. All three different light-matter interactions that we know now, viz., absorption, spontaneous emission, and stimulated emission, are represented schematically in Fig. 5.1. As seen, the ground state atom absorbs this resonant photon[1] and readily climbs to the excited state. This atom, as we know, cannot stay excited indefinitely and spontaneously returns back to the ground state after a finite amount of time, releasing the energy of excitation as another photon. In the third process, called stimulated emission, the excited atom instantaneously relaxes back to the ground state, in the presence of a resonant photon, emitting another photon identical to it in all respect. These two unidirectional photons moving in unison, we would tend to believe, can draw two more by setting up stimulated emissions in two more excited atoms, thus growing into four. This, therefore, should result in a collection of increasingly many identical photons through stimulation in further and even further excited atoms. The end result will thus be a whole bunch of identical unidirectional photons establishing precisely a beam of light. If this were true, then the laser would have seen the light of the day more than a century before it actually was born, parallel with the invention of fluorescent light [20] by the German glass blower Heinrich Geissler (1814–1879) in 1856. Surely there is a glitch somewhere; else conceptualization of stimulated emission by Einstein would not have predated the development of a laser by over four decades!

With a little introspection, the truth readily emerges. We erred, first, by not considering the possibility of losing photons through the process of absorption. Surely, a photon gives birth to a second identical photon when it stimulates an excited atom, but the photon can also be lost through absorption if it chances upon a ground state atom instead. In summary, a stimulated emission adds one photon into the system, while the process of absorption removes one photon from it. Obviously, therefore, the number of photons can multiply if only the number of excited atoms exceeds their number in the ground state. In a state of thermal equilibrium, as we know from Boltzmann's law, the population of an energy level exponentially decreases with increasing energy. Considering now the case of the fluorescent tube light, the number of spontaneously emitted photons travelling along the length of the

---

[1]The photon here is a resonant photon as its energy matches exactly the difference of energy between the excited and ground states.

Fig. 5.1 Schematic illustrations of the processes of absorption and, spontaneous and stimulated emissions. A ground state atom can climb to the excited state by absorbing a resonant photon. The excited state is short-lived, and the atom eventually returns back to the ground state spontaneously releasing the energy of excitation as another photon. The photon can be emitted in any random direction. In the presence of another resonant photon, the excited atom is stimulated to instantaneously return back to the ground state by emitting a photon that is identical to the stimulating photon in all respects

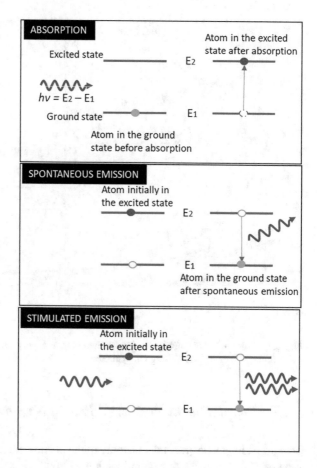

tube can never multiply as they will cause less stimulated emission and undergo more absorption. This prevents amplification of light in a fluorescent tube, and the dominating role of spontaneous emission here letting emitted photons spread in all directions makes it an ideal viewing source instead. For a bunch of spontaneously emitted photons, travelling along the length of the tube, to develop into a beam of light, they must be able to manufacture more photons through stimulation than perishing through absorption. This can happen only when the excited atoms out-number them in the ground state: a situation that can most accurately be labeled population inversion. A population inverted state, as we shall see in the following section, can be best described as a case where Boltzmann's population distribution law is overturned in complete violation of the rule of nature.

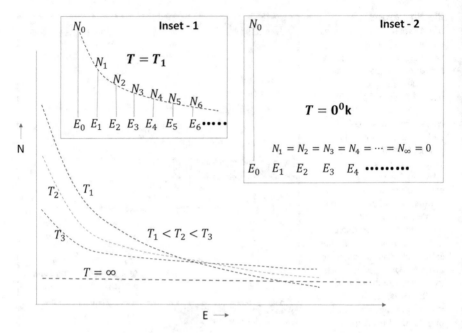

Fig. 5.2 Distribution of population among the discrete energy levels lying between zero (E_0) and infinite (E_∞) energy at varying temperatures

5.2.1 Population Inversion: A State of Negative Temperature

Toward gaining a deeper understanding of the phenomenon of population inversion, a prerequisite to achieving amplification of light, it is imperative to revisit the Boltzmann law introduced earlier in Chap. 3. Boltzmann's equation governs the distribution of the population N_j among the permitted energy levels E_j, and the same is schematically illustrated in Fig. 5.2 for temperatures ranging from absolute zero to infinity. Inset-1 of this figure shows population N_0 of ground energy level E_0 and also populations N_1, N_2, N_3, N_4, and so on of the correspondingly excited energy states one above the other, viz., E_1, E_2, E_3, E_4, and so on. The population points have been joined with a dashed line only to impart an impression of gradual reduction of population with increasing energy which asymptotically approaches a zero value at the level with infinite energy. This line has no other practical significance as the intermediate space between any two adjacent discrete energy states is devoid of any population. The central point of this figure is to elucidate the role of temperature on the distribution of the population. The asymptotic reduction of the population with progressively increasing energy is schematically illustrated here for three different finite temperatures T_1, T_2, and T_3 where $T_3 > T_2 > T_1$.

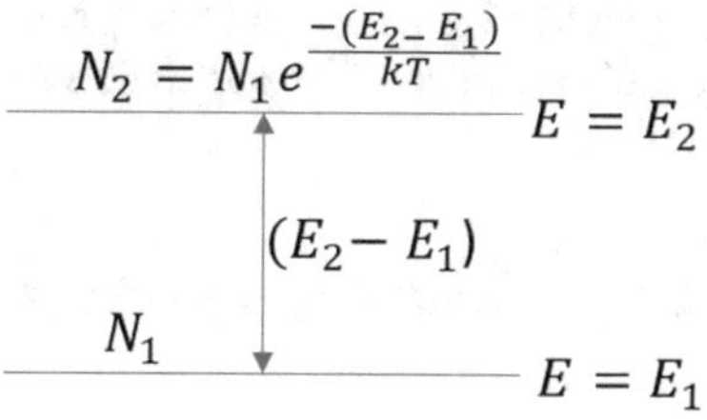

Fig. 5.3 Distribution of
population between two
energy levels satisfying
Boltzmann's law

$$N_2 = N_1 e^{\frac{-(E_2 - E_1)}{kT}}$$

_____ $E = E_2$

$(E_2 - E_1)$

N_1

_____ $E = E_1$

$$N_j = N_0 e^{\frac{-E_j}{kT}} \tag{5.1}$$

Unmistakably, with rising temperature, the higher energy states become increasingly more populated at the expense of populations of ground and lower energy levels. Significantly though, however high the temperature may be, at no point will any energy level have population exceeding that of any level beneath it. Even in the limiting case of an infinite temperature, they can at most be equal as has also been depicted in this figure. Just as the level with higher energy gets increasingly more populated with increasing temperature, exactly the opposite happens for reducing temperature; the levels with lower energy get increasingly more populated at the expense of population of higher energy levels. In the limiting case of the temperature dropping down to absolute zero, the entire population of the system collapses to the ground state. This situation is illustrated in inset-2 of Fig. 5.2. Obviously then, realization of population inversion is not possible for any temperature lying between zero and infinity.

To formulate a Boltzmann distribution-based interpretation of population inversion, it is imperative to apply this theorem to the distribution of the population between two energy levels as depicted in Fig. 5.3. Clearly populations N_2 and N_1, of energy levels E_2 and E_1, respectively, will be connected through the following equation:

$$N_2 = N_1 e^{\frac{-(E_2 - E_1)}{kT}} \tag{5.2}$$

A little inspection of this equation establishes beyond doubt the fact that N_2 can exceed N_1 if, and only if, T becomes negative. This is equivalent to saying that a state with population inversion is basically a negative temperature state. In reality it is a state of thermal nonequilibrium as it violates Boltzmann's distribution. Such an unstable situation, which defies the rule of nature, is always short-lived. There will be a rapid redistribution of populations within the constituent energy levels, reestablishing a new thermal equilibrium.

Such a negative temperature condition, which represents a state of population inversion, cannot obviously be realized by elevating the system temperature, nor can it be achieved by cooling the medium, the lower limit of which is absolute zero. The only way to raise the population of a particular energy state beyond the permitted value, resulting in an inversion of population, will be by selectively transferring population from a lower energy state into it. Some of the techniques that have gained popularity in creating population inversion are described in the following section.

5.2.2   Population Inversion: Methods of Creation

It is worthwhile to consider two energy levels, one ground (E_0) and the other excited (E_1), between which population has to be inverted (Fig. 5.4). A majority of the members of the system will be residing in the ground state to begin with. The obvious requirement is to transfer a fraction of them to the excited state so as to make it more populous than the ground state, a crucial step for the making of a laser. This process of transferring population is, in the common parlance, termed pumping of the laser.

5.2.2.1   Optical Pumping

As the name suggests, this involves employing ordinary light to manufacture coherent light and is generally utilized in pumping solid and liquid lasers. Optically pumped gas lasers, however, are also not uncommon. Here, the lasing medium is required to be shone by a stream of photons of energy matching exactly the difference of energy between the excited (E_1) and ground (E_0) states (Fig. 5.4a). The species (atoms, molecules, or ions) residing in the ground state, upon absorption of these resonant photons, will make quantum leaps (represented by arrows in Fig. 5.4b) to the excited state. Augmenting the number of these shining photons appropriately should result in striking a population inversion state (Fig. 5.4c). A two-level system is considered here only for rendering a simplistic view of the process, and we would soon learn in a latter section of this chapter that the participation of at least one more energy level is mandatory for the realization of population inversion.

The light to optically pump the lasing medium is often derived from a flashlamp that basically is a fluorescent light source. The most common type of flashlamps normally has a helical geometry that spirals around the active medium as shown in Fig. 5.5. This arrangement allows light from all sides to shine on the lasing medium.

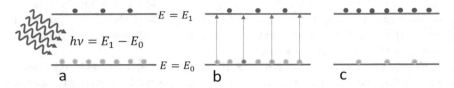

Fig. 5.4 Illustration of optical pumping in a two-level system. Resonant photons, upon shining onto the medium that is at a thermal equilibrium (**a**), are absorbed by the ground state species (**b**) and, in turn, result in the creation of population inversion (**c**), rendering the medium into a state of thermal nonequilibrium

Fig. 5.5 Schematic illustration of optical pumping of a solid-state laser by means of a spring-shaped flashlamp. The cylindrical enclosure to reflect the flashlamp light that moves outwardly is also shown here

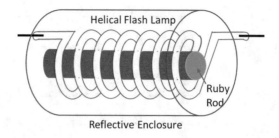

The entire assembly is often placed inside a hollow cylindrical enclosure possessing a reflective inner surface. This enhances the pumping efficiency as the light moving outwardly from the flashlamp can now be put back into the lasing medium through reflection. Although flashlamp pumping usually finds application in the case of solid-state lasers, its use in liquid lasers is also not uncommon. The very first laser, viz., a ruby laser, that Maiman operated in 1960 made use of a helical flashlamp wrapped around the ruby rod. Optical pumping of gas lasers, on the other hand, is usually accomplished by deriving coherent light from a second laser. The optical pumping schemes that vary from laser to laser will be described in greater detail in Chap. 10 dealing with specific laser systems.

5.2.2.2   Electrical Pumping

In a majority of the gas lasers, excitation of the species from the ground to the higher energy level is achieved by taking advantage of the avalanche of electrons in a high voltage discharge much the same way as in the case of a fluorescent light. The active species, depending upon the laser, can be either an atom, a molecule, or an ion. Let us consider it, at this point, to be an atom. As the electrons move in the discharge, they gain velocity and thus acquire kinetic energy. These energetic electrons inevitably will undergo collisions with the numerous atoms present in the discharge volume, of which the majority are in the ground state to begin with (Fig. 5.6A). The inelastic collision of an electron with a ground state (E_0) atom can excite it to a higher energy level (E_1), and the electron, having transferred a part of its energy to the atom, now moves with a lower KE. This process is captured in the following equation:

$$\text{Atom}_{E_0} + e(\text{KE}) = \text{Atom}_{E_1} + e(\text{Less KE}) \tag{5.3}$$

Under favorable conditions, the number of atoms in the excited state may outnumber those present in the ground state (Fig. 5.6B) paving the way for the amplification of light. The experimental arrangement to effect electrical pumping for a continuously working laser differs significantly from that meant for pulsed lasers and will be addressed in greater detail when we study specific gas laser systems in Chaps. 10 and 11.

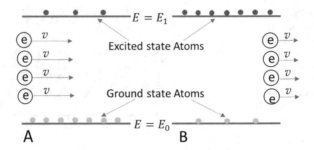

Fig. 5.6 Schematic illustration of creation of population inversion wherein the electrons in a high voltage discharge collide with the atoms in the ground state and transfer part of their KE to the atoms (**A**). As a result of these inelastic collisions, some of the ground state atoms get excited to the higher energy level much the same way as in the case of a fluorescent light. The electrons after the collisions move with reduced KE (**B**)

5.2.2.3 Other Pumping Schemes

There exist a variety of pumping schemes that do not have general applications but are suitable for creating population inversion only in some specific laser systems. Population inversion is automatically created when a hot gas is made to rapidly expand into a cold region. A gas dynamic laser operates by taking advantage of this phenomenon. In certain chemical reactions, the product is formed in an excited state, thereby intrinsically establishing the condition of population inversion. The laser that operates by taking advantage of this fact is known as a chemical laser. Certain rare halide molecules exist only in an excited state, meaning that the ground state population here is always zero. The formation of such molecules thus spontaneously results in the inversion of population, and the lasers that operate on this principle are usually referred to as the excimer laser. All these different but nonetheless specific classes of lasers will be addressed in more detail later in this book. It is worthwhile to mention here just for the sake of completeness that a laser has also been pumped by utilizing the highly energized nuclear particles emanating from the explosion of a nuclear bomb. Such a laser, suitable for military application, can give out an intense pulse of laser light before the shock wave, the post effect of the explosion, inevitably destroys it.

5.2.3 *Population Inversion Vis-à-Vis Number of Participating Energy Levels in the Process of Lasing*

In the examples of preceding section, we considered, for the sake of simplicity, the process of creation of population inversion between the ground and the excited states without involving any other energy levels. This section begins by addressing the insurmountable hurdles posed by a two-level system toward achieving population inversion and, in turn, its impracticality to allow amplification of light. In a real laser, the involvement of at least one more level only allows the creation of population inversion and, in turn, lasing. Physical insight into overcoming the challenges of a two-level system by going for a three-level configuration also forms a major part of this section. The section ends with an explanation as to how a four-level laser system can be construed as an ideal laser. Optical pumping, capable of readily providing resonant photons, prerequisite for the onset of absorption and stimulated emission processes, will be made use of as we examine here the pros and cons of the process of lasing involving two, three, and four energy levels.

5.2.3.1   Two-Level System

We begin by considering two atomic energy levels, one ground (E_0) and the other excited (E_1), into which the distribution of the population is in a thermal equilibrium. The majority of the atoms are, to begin with, therefore, residing in the ground state, and the excited state is barely populated. The medium is now shone upon by a beam of light comprising resonant photons far exceeding the total number of atoms making up the ensemble. We now know that a photon upon interaction with a ground state atom will be absorbed and the atom, in turn, will be excited. The interaction of the photon with an excited state atom, on the other hand, will cause a stimulated emission creating another identical photon, and at the same time, the atom drops down to the ground state. In the beginning, as the majority of the atoms are in the ground state, more photons will be absorbed from the incident beam than will be added to it due to the stimulated emission. The corresponding attenuation in its intensity is reflected in Fig. 5.7A as a reduction in the beam width. As this continues, the excited state population will gradually build up at the expense of that in the ground state. This would, in turn, cause a steady rise in the rate of stimulation and a corresponding fall in the absorption. This fact is schematically depicted in Fig. 5.7B as an increase in the number of atoms in the excited state in conjunction with a rise in the intensity of the transmitted beam. A situation would therefore inevitably emerge where the excited state population equals that at the ground state and will remain so forever. Beyond this point, the incident beam of photons will continue to interact evenly with the atoms in the two states and thus would emerge through the medium unscathed (Fig. 5.7C) leading inescapably to a stalemate-like situation. As the number of atoms in the ground and the excited states is the same, the probability of interaction of the incident beam of light with either of the energy states also becomes equal. At any instant if one photon from the incident beam is lost due to absorption, the process of stimulated emission adds a photon to it at that instant. Undoubtedly, therefore, it would be impossible to create population inversion in a two level-based system. A laser thus can never be conceptualized on a two-level scheme, let alone its operation.

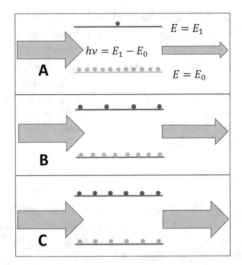

Fig. 5.7 When a two-level medium is shone by a beam of resonant photons, the ground state (E_0) atoms will begin absorbing the photons from the incident beam. The corresponding reduction in the intensity of the beam, as it passes through the medium, is reflected here in its reduced width (**A**). The number of atoms in the excited state (E_1) will thus gradually rise at the expense of their number in the ground state. This results in increased rate of stimulation and, in turn, an increase in the intensity of the transmitted beam (**B**). A situation will inevitably arise where the two states become equally populated and the pump beam passes through the medium intact (**C**)

5.2.3.2 Three-Level System

This two-level impasse can be readily overcome by involving the service of a third energy level in the process of lasing. The third energy level, called the pump level E_2, is usually located slightly above the second level that serves as the upper level E_1 for the lasing. Such a three-level scheme and its operation are schematically described in the traces of Fig. 5.8. The lower laser level is the ground level E_0, which we shall also consider as the source level, meaning that the majority of the population initially resides here. For an easy understanding of the working of this three-level scheme, we shall assume that both the pump and the upper levels are empty to begin with (trace A). The medium is now shone upon by a beam of resonant photons that fits exactly within the energy gap between the pump and the ground levels (trace B). The ground state atoms will begin absorbing these photons and climb up to the pump level. The residence time of the atoms at the pump level is extremely short, and, upon arrival here from the ground level, they rapidly relax down to the upper level. The energy is usually released as heat and not light in this nonradiative relaxation process. The rapid relaxation prevents the incident photons from catching these atoms at the pump level and drives them back to the ground level through stimulated emissions unlike in the two-level case. Thus, in a way, the incident light basically transfers the atoms from the ground state to the upper level

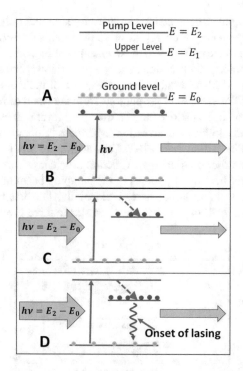

Fig. 5.8 Schematic representation of a three-level scheme where the pump and the upper laser levels are decoupled (**A**). When shone by a resonant beam of photons, the atoms in the ground state, upon absorption of the photons, begin to appear in the pump level (**B**). The pump level is connected to the upper laser level by a very fast nonradiative process (indicated by the dashed arrow), and the atoms from the pump level rapidly appear in the upper level (**C**). The upper level is a metastable state, and so the population gradually builds up here and eventually exceeds that at the ground level causing the onset of lasing (**D**)

via the pump level (trace C). The incident photons fail to interact with the atoms at the upper level as their energy does not equal the difference in energy between the upper and the ground level. The upper level is usually a metastable state[2]; therefore, the atoms reside here for a relatively longer period and do not immediately sponta-neously decay back to the ground state. This allows the gradual accumulation of the population at the upper level giving rise to the prospect of inverting the population between the upper and the ground levels (trace D).

The realization of population inversion automatically results in the onset of lasing as the spontaneously emitted photons, while travelling through this inverted medium, will cause more stimulated emission than their loss through absorption. Clearly a three-level laser suffers from the disadvantage that the lower laser level is also the source or ground level where the entire population resides in the beginning.

---

[2] A metastable state is an excited energy level of a microscopic particle like an atom/molecule that has a lifetime much longer than that of typical excited states.

The condition of population inversion can, therefore, be met only at the expense of an exorbitantly large value of energy as more than half of this huge population is required to be lifted to the upper level. In the example of Fig. 5.8, all 11 atoms, which make up the ensemble, initially reside in the ground or lower laser level. Population inversion could be established only when at least six of them are transferred to the upper laser level at the expense of six photons from the beam of incident light. That is not all; stimulation of just a lone atom at the upper level will bring it down to the lower level, instantly killing the inversion. Maintaining population inversion in a three-level laser is indeed an expensive affair, and, quite predictably, this class of lasers suffers from rather poor energy extraction efficiency. This is reflected in the fact that a significant fraction of photons of the pumping beam of light is consumed in initiating and sustaining lasing here. Incidentally, the first laser, a ruby laser operated by Maiman in 1960 [8], was based on such a three-level scheme.

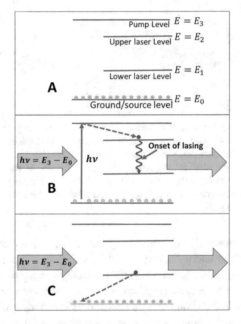

Fig. 5.9 Schematic of a four-level laser where both upper and lower laser levels are, respectively, decoupled from the pump and ground levels. The entire population resides initially in the ground level (**A**). The atoms excited to the pump level by the incident photons rapidly arrive in the upper laser level. Population inversion and, in turn, lasing can be achieved with relative ease as the lower laser level is empty (**B**). Extremely fast relaxation of population from the lower to the ground level helps maintain population inversion and uninterrupted lasing (**C**). The rapid relaxations between pump level to upper laser level and lower laser level to ground are indicated by the dashed arrows

5.2.3.3   Four-Level System

The inherent limitation of a three-level laser can be eliminated in a straightforward manner by going for a four energy-level scheme. This allows separation of the lower laser level from the ground level. Removal of the coupling between the lower laser level and the ground or source level, which basically mires the operation of a three-level laser, makes all the differences. The schematic of a four-level laser is depicted in Fig. 5.9. As seen, both the upper and lower laser levels are separated here from the pump and source levels, respectively. The lower laser level is located so far away from the source level that it is practically devoid of possessing any thermal population (**trace A**). The incident beam of photons, as in the case of the three-level configuration, transfers the population from the source to the pump level that rapidly relaxes to the upper laser level. The empty lower laser level ensures the realization of population inversion even if only a lone atom has been transferred to the upper level via the pump level triggering the onset of lasing (**trace B**). Furthermore, the rapid relaxation of the lower-level population to the ground level ensures that the population arriving here as an upshot of lasing are immediately drained to the ground and, in turn, their nonstop participation in the process of lasing (**trace C**). The inherent ability of the four-level scheme to preserve population inversion with relative ease makes this class of lasers the most ideal. It is not surprising that they intrinsically offer high energy extraction efficiency. Today most of the common lasers operate on a four energy-level scheme, and we will learn more about them in the latter chapters dealing with specific laser systems.

Chapter 6
Cavity and Its Bearing on the Operation of Lasers

6.1 Introduction

We know by now that the key to amplifying light is to establish the process of stimulated emission in a medium in which the population has been inverted between two energy levels. We also have learned in the preceding chapter about some common techniques to effect population inversion and the ideal energy level configuration where it can be created and sustained with the minimum effort. In a fluorescent tube, excitation is indeed created but not inversion, which explains why the laser did not come into existence parallel with the invention of fluorescent light in 1856. The obvious question that arises here is "would it have been possible, if a mechanism conducive to create and maintain population inversion were ensured, to obtain an amplified beam of light from this device?" Let us dig a little deeper in pursuit of a plausible answer to this question. The operation of a fluorescent light has been elaborated in Chap. 4. In brief, a high voltage glow discharge is struck between two electrodes fixed at the two ends of a cylindrical glass tube that contains a rarefied atomic gas. The discharge manifests through the ionization of a fraction of the neutral gas atoms A into cations A^+, while another fraction of it is driven into the excited state A^*. Electrons e equal to the number of ions A^+ will also be inevitably present. We, of course, begin with the presumption that a condition of population inversion exists inside the discharge tube, meaning the number of atoms in the excited state, namely A^* outnumber that in the ground or unexcited state. The situation is schematically illustrated in Fig. 6.1. As the excited atoms begin dropping spontaneously to the ground state inside the tube, photons will be emitted in all possible directions. To simplify the interpretation, we catch one such photon at the extreme left end of the tube, indicated as a red curly arrow in this figure, that is traveling along its axis to the other end. The existence of population inversion ensures that upon reaching the right end, it will grow in numbers. The obvious question is by how much? Light travels at an enormous speed, and the gas atoms,

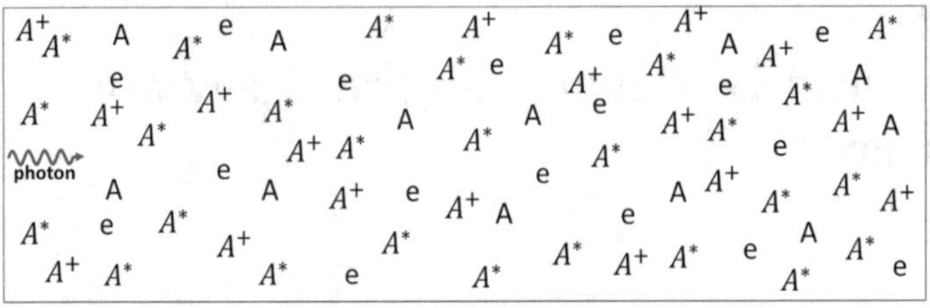

Fig. 6.1 Under the condition of population inversion, the number of excited atoms (A^*) will exceed that in the ground state (A). Electrons (e) and ions (A^+) will also be present in the discharge in equal numbers

which occupy only a tiny fraction of the volume inside the tube, move in a zigzag fashion. In the split second that the photon will take to traverse the typical length of a fluorescent tube, it obviously cannot grow to the extent of developing into a beam of light to produce an effect visible to the naked eye. The growth understandably will be even correspondingly less for photons originating at points increasingly closer to the right end. The same also holds true for all the photons moving along the axis from the right end to the left. We are, therefore, now in a position to answer the question posed earlier: The realization of population inversion, in a fluorescent tube light, will definitely cause amplification of spontaneously emitted photons in either direction but cannot, alone, result in the formation of a beam of light, typical of lasers.

The most obvious way to increase the possibility of the interaction of photons and atoms, preferably the excited ones, is to stack them up more closely. However, this will work only up to a certain extent as increasing the gas pressure will allow electrons progressively less path of travel between successive collisions. This understandably will restrict the K.E. of the electrons to a lower value and, in turn, adversely affect both the sustenance of the discharge and the process of creation of population inversion. (For refreshing the mind on the role electrons play here, the readers are advised to take another look at Sect. 4.2). The only other way to augment the photon-atom interaction will be to extend the length of the interaction. As it would be impractical to extend the length of the tube to a great extent, making the photons traverse the population inverted medium multiple times seems to be a workable solution (Fig. 6.2). Placing a plane reflector at one end of the tube allows the photons to traverse the length only twice (Fig. 6.2a). Placing a second plane reflector on the other end parallel to the first will make the photons travel through the medium time and again (Fig. 6.2b) allowing them to grow astronomically. This, however, can happen if and only if the two reflectors are absolutely parallel to each other. No tolerance on parallelism is permissible, for even an insignificant departure from parallelism would make the light ray simply fly away as schematically illustrated in Fig. 6.3. An optical cavity is formed when two such reflectors are placed parallel to each other at a given separation. Imparting concavity to at least one of the reflectors will add stability to the cavity offsetting its critical dependence on the degree of parallelism. Aspects of a cavity including its stability criterion and ability

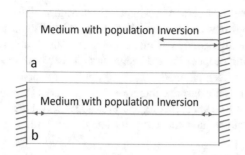

Fig. 6.2 Augmentation of the photon-atom interaction is possible by making the photons traverse through the length of the population inverted medium multiple times. A reflector placed at one end of the fluorescent tube will allow the photons to travel through the medium only twice (trace a). Two parallel reflectors placed at the two ends will make light pass through the medium repeatedly (trace b)

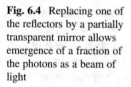

Fig. 6.3 The left reflector is normal to the axis of the tube, while the reflector on the right is slightly off. An axial ray of light, as it travels to and fro between the reflectors, gradually shifts laterally before eventually flying away

Fig. 6.4 Replacing one of the reflectors by a partially transparent mirror allows emergence of a fraction of the photons as a beam of light

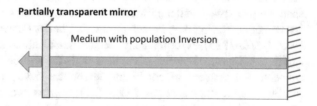

to enrich the purity of the color of light will be elaborated in the following sections. A total reflector, however, does not let any light leak through. Light instead stays confined between two such reflectors, and such a device thus would not as such allow the realization of a laser beam.[1] Obviously, only a leaky reflector will allow the emergence of a fraction of these trapped photons as a beam of light. This is where a partially transparent mirror, described in Chap. 2 (Sect. 2.5.1), comes handy. A feasible scheme to extract a beam of light from a medium with population inversion is schematically illustrated in Fig. 6.4. However, the extraction of light from the cavity will be possible only when, as we shall see in the following section, the level

---

[1] We shall learn in a future chapter that the ability of such a cavity to store light in its entirety can be exploited to manufacture ultrashort pulses in the operation of a certain class of lasers.

of leakage is kept below a certain maximum. Furthermore, optimization of the cavity leakage greatly improves both the quantity and quality of light that it gives out. All these facts will be enlightened in greater detail in the latter sections of this chapter. In addition to providing deeper physical insight into the behavior displayed by a cavity, this chapter has also been planned to bring to the fore the pivotal role it plays in the operation of a laser. Little surprise therefore that the Nobel Prize for the invention of laser was awarded to Townes, Basov, and Prokhorov who independently conceived the concept of a cavity and theorized its integration with a population inverted medium.

6.2   Laser: An Amalgamation of Cavity and Population Inversion

We know that the spontaneously emitted photons, upon passing through a population inverted medium, will grow in numbers due to the dominance of stimulated emission over absorption. Such a medium that is capable of amplifying light is also called a gain or active medium. Given the typical length possible for such an active medium, the intensity of the amplified light, having traversed through it only once, cannot, however, attain a level readily distinguishable from the light given out by a normal source.[2] There must be a way to enhance the intensity by trapping this light and then making it pass through the gain medium over and over again. A provision to tap a fraction of these trapped photons out will yield a beam of light preserving all the characteristics of stimulated emission. That light can be stored in the intermediate space between two parallel mirrors has been known almost from the beginning of the last century when in 1899, French physicists Charles Fabry (1867–1945) and Alfred Perot (1862–1925) experimentally demonstrated the concept with a tiny optical cavity [9]. Charles Townes is credited with the distinction of being the first to conceive the concept of combining an optical cavity and population inverted medium for the realization of a laser [21]. To this end, the quality factor of the cavity, which basically governs its leakage, plays a critical role. This indeed is quite intriguing, and it is worthwhile in this context to carry out an investigative experiment schematically described in Fig. 6.5. The lasing medium, which is continuously pumped to create and maintain population inversion, is enclosed between two parallel mirrors (upper trace). One of the mirrors is 100% reflective, while the other is partially transparent, and the reflectivity R of which can be discretely varied from 10% to 100%. The output of the laser obviously emerges through this semi-transparent mirror. The intensity (I) of the laser beam is monitored by a light detector as a function of time for five different values of R, viz., *10%, 50%, 80%, 90%,* and

---

[2]Exception to this statement is very high gain systems where amplified spontaneous emission may exhibit some of the characteristics of a laser like its monochromaticity; a case in point here is Maiman's famed experiment.

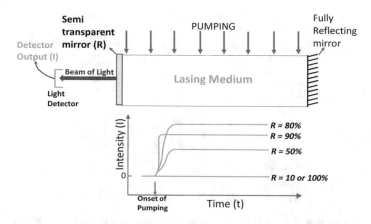

Fig. 6.5 Critical dependence of the output intensity (*I*) of a typical laser on the reflectivity (*R*) of the output mirror as a function of time (*t*) is depicted (lower trace). The laser is being continuously pumped to create and maintain population inversion (upper trace). The instant of onset of pumping is indicated by an arrow on the time axis

100%, and the dependence is schematically depicted in this figure (lower trace). Obviously, the laser would fail to produce any output until the medium is pumped to create a population inversion. Surprisingly, however, the laser continues to give zero output even after the onset of pumping when *R* is *10%*. Once the condition of population inversion is established, the dominating role played by the stimulated emission over absorption, as we know, should result in the amplification of sponta- neously emitted photons. Ninety percent of these amplified photons are expected to emerge through this 10% reflective mirror. Intriguingly, though, even after a long wait from the instant of the onset of pumping, no light, however, is traceable by the detector. Population inversion does exist and yet no amplification of light! Where lies the catch then? Let us take another look at the population inverted medium of Fig. 6.5 in an attempt to crack the puzzle. We tag a lone spontaneously emitted photon at the extreme left end of the population inverted medium heading directly toward the total reflector located on the other end (Fig. 6.6a). This photon grows in number as it reaches the right end of the medium. The mirror at this end will reflect all these identical photons back, and during the reverse journey, their number will continue to grow. Let us assume that upon reaching the 10% reflective mirror, located on the left end of the active medium, the lone photon has been able to draw seven more (Fig. 6.6b). This mirror will allow 90% of these eight photons to pass through while reflecting 10% back that amounts to even less than one photon! In a nutshell, this means that amplification of light is not possible in this cavity that is 90% leaky, no matter how many to and fro journeys it makes. Unmistakably, the gain of photons due to the stimulated emission is being outdone here by the combined loss of photons due to absorption and the bulk transmission loss through this leaky mirror. Raising the population inversion, which, in turn, raises the gain due to stimulated emission, and slashing the transmission loss of the cavity are

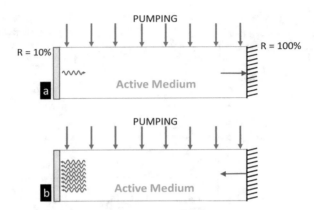

Fig. 6.6 Unless gain due to stimulated emission cannot overcome the combined losses due to absorption and cavity leakage, amplification of light is not possible. A lone spontaneously emitted photon at the extreme left end of the active medium heads directly toward the total reflector (trace a). As it returns back to the starting point after completion of a to and fro journey, it draws seven more identical photons (trace b). 90% of these 8 photons will pass out through the 10% reflective mirror

obviously the two plausible ways to overcome this problem. In this experiment, we follow the second route, i.e., making the mirror less leaky or equivalently more reflective. We therefore change the reflectivity of the semitransparent mirror now to 50% and study the output of the light detector in time as before. It is seen that immediately following the commencement of pumping, the detector senses light and its intensity gradually builds up in time until a stable value is reached in the steady state.[3] This behavior is obvious: As a consequence of the reduction of the transmission loss from *90%* to *50%*, the number of photons in the cavity can now obviously grow in time and soon attain an equilibrium value as the pumping is maintained at a constant level. This intracavity dynamics of photons over time is also reflected in the intensity of the emerging light as recorded by the detector and displayed in Fig. 6.5. We next raise the reflectivity of the semitransparent mirror to 80%, and consequently the transmission loss reduces further and assumes a value of *20%*. A higher fraction of photons that remains inside the cavity after every round trip not only results in accelerating the growth of the photons but also significantly raises their number in the steady state. The light detector now accordingly records a faster rate of buildup of intensity reaching an equilibrium value far exceeding that in the previous case. It may thus seem that the rate of buildup of laser intensity and its steady state value will monotonously increase with increasing reflectivity of the semitransparent mirror. However, that is not quite true as would be revealed in the next experiment carried out with a *90%* reflective semitransparent mirror. The cavity will retain *90%* of the photons created in every round trip compared to the *80%* retention in the previous case. The presence of more photons will further accelerate their production in time inside the cavity. This fact is reflected in the even faster buildup of the intensity of

---

[3] Steady state represents a state where the properties of the system do not change with time.

Fig. 6.7 The power given out by a laser exhibits a maximal behavior with the reflectivity of its output mirror. There also exists a threshold reflectivity for the onset of lasing. The second cavity mirror is fully reflective

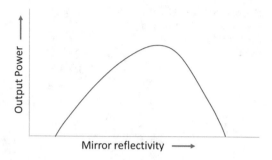

light given out by the laser in this case. The output intensity of the laser, however, settles to an equilibrium value lower than the case with *80%* reflective mirror. Although the number of intracavity photons in the steady state is now higher, *10%* leakage of them cannot beat the numbers that could leak out of the cavity in the previous case. Thus, the laser output will exhibit maximal behavior with increasing reflectivity of the semitransparent mirror. We finally make the cavity nonleaky by raising the reflectivity of the semitransparent mirror to *100%*. As the cavity now will retain all the photons being produced in every round trip, their number will build up at the fastest rate and will continue to rise as seemingly there will not be any steady state in the absence of any cavity loss. With no possibility of leakage of any photon from this cavity, the laser will not yield any output.

It should be noted here that the sole purpose of engaging the readers with this illuminating experiment is to bring afore the centrality of the cavity in the realization of a laser. Further, there is no inviolability in the specific reflectivity of the semitransparent mirrors chosen for this experiment. The experiment reveals that for a given length of the active medium and the strength of population inversion, there exists a threshold reflectivity for the onset of lasing and that the output of the laser exhibits a maximal behavior with the cavity reflectivity. Figure 6.7 depicts the dependence of output power of a laser as a function of the reflectivity of its cavity; the threshold reflectivity and the reflectivity at the maximal point will undoubtedly vary from laser to laser. Although the interpretation of these concepts has been very qualitative thus far, an attempt will be made to present an analytical formulation, whenever possible, without compromising the basic objective of this book.

6.3 The Cavity: A Deeper Insight

A cavity not only has made possible the realization of a laser, as we just saw in the preceding section, but also has gifted it all its prized attributes. It would be beneficial at this point to gain a qualitative, but nevertheless deeper, insight on certain unique features of a cavity that have strong bearing in the operation of a laser. Let us begin by considering a typical Fabry-Perot cavity comprising two parallel mirrors each of reflectivity ~99% and separated by a distance *l* (Fig. 6.8). Let a beam of light of power *1 W* shines upon it. The front mirror, being 99% reflective, will allow only *1%*

Fig. 6.8 A typical cavity usually reflects back almost the entire light incident upon it

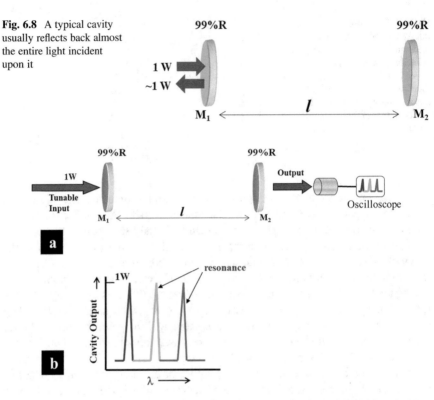

Fig. 6.9 Measurement of transmission of a cavity as a function of the wavelength of the incident light. A typical experimental layout for such a measurement is shown in trace a. Input to the cavity is derived from a tunable source. Transmission as a function of wavelength is monitored by placing a detector on the other side of the cavity. Recording of the cavity transmission as a function of wavelength of the incident light is depicted in trace b

of this input, i.e., *10 mW*, to get inside the cavity. Of this 10 mW, the rear mirror reflects *99%* back transmitting therefore just about *100 µW* out. Thus, it is obvious that the cavity transmits merely an insignificant fraction of the light incident on it. Fortunately, this is not always true; else, the cavity would have lost all its importance in the context of a laser. A simple experiment can be performed to make the cavity reveal its deepest secrets, and the same is depicted in the trace of Fig. 6.9. For this, we derive the input beam of light from a tunable source that is capable of giving out light of continuously changing wavelength. The transmission of the cavity is monitored as a function of the wavelength of the incident light by placing a detector on its other side (Fig. 6.9a) and the same is recorded in Fig. 6.9b. As expected, for most of the wavelengths, the cavity hardly transmits any light. Intriguingly, however, at certain discrete wavelengths, the cavity is seen to transmit almost whatever is incident on it. For these wavelengths, the cavity is said to be on a resonance, and the corresponding wavelengths are called resonant wavelengths. Thus, a resonant cavity is one that transmits the entire light that shines upon it, while a nonresonant cavity reflects it back almost entirely (Fig. 6.10). On closer

Fig. 6.10 A resonant cavity
transmits entire light, while
a nonresonant cavity reflects
the incident light wholly

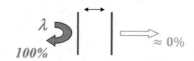

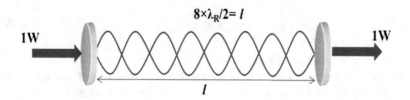

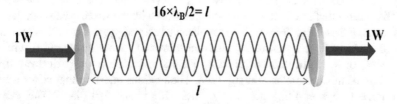

Fig. 6.11 Illustration of two examples of resonance occurring in a cavity for two different wavelengths. For the longer wavelength λ_R, the resonance condition is satisfied for an integer value of 8 (top trace), while for the shorter wavelength λ_B, the resonance condition is satisfied for an integer value of 16 (bottom). In reality, however, the value of n is much higher. For visible light, this value of n can run into several millions for a typical cavity of 1 m length

examination, it would be seen that only those wavelengths, the half of which times an integer fits exactly inside the cavity, are actually resonant to the cavity. This fact is schematically illustrated in Fig. 6.11 for two different wavelengths. The resonant equation can be mathematically expressed as

$$n \times \lambda_n/2 = l \tag{6.1}$$

where n is an integer, l is the length of the cavity, and λ_n is the wavelength corresponding to the nth integer. In terms of frequency ν_n, this equation can be rewritten as

Fig. 6.12 Schematic illustration of electrons orbiting in circles with the nucleus at the center. An electron can only reside in those orbits into which its de Broglie wavelength fits an integer number of times

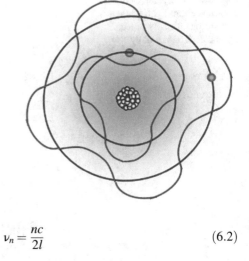

$$v_n = \frac{nc}{2l} \tag{6.2}$$

where "c" is the velocity of light.

These discrete frequencies of light for which it can oscillate back and forth inside a cavity without decay are called cavity modes. An obvious analogy is Bohr's orbits of electrons. It is well known that electrons can revolve only in those orbits without decaying into which its de Broglie wavelength fits an integer number of times (Fig. 6.12). Another example is the case of a plucked string where only those wavelengths survive that also bear a similar relationship with the spacing between the two fixed ends of the string. This finding is applicable to a variety of musical instruments.

Although we have a resonance, the front mirror of the cavity should still reflect back 99% of the light incident upon it, and the calculation we performed at the beginning of this section should still hold. The obvious question that arises here is how come then everything gets to the other side in the first place? This can be explained in the following manner.

The tiny fraction of the input light, viz., 10 mW, that enters the cavity oscillates back and forth inside it and causes the intracavity photon flux to gradually rise as the input beam is continuous. This continues until the intracavity power attains an equilibrium value of 100 W. At this instant, *1% of this intracavity power, viz., 1 W,* is transmitted out of the rear mirror. Thus, in equilibrium, for *1 W* input, the cavity also yields an output of *1 W* maintaining all the while an intracavity power[4] at the level of *100 W*. This situation is illustrated in the traces of Fig. 6.13.

Armed with this knowledge, we can now explain the transmission of the cavity as a function of the wavelength of the input light captured in Fig. 6.9b. Whether light of

---

[4]While the front mirror reflects 99% of the power incident on it, for an equilibrium power of 100 W inside the cavity, it also transmits an equal amount that interferes destructively with the reflected component leading to a complete coupling of the incident power into the cavity under a resonant condition.

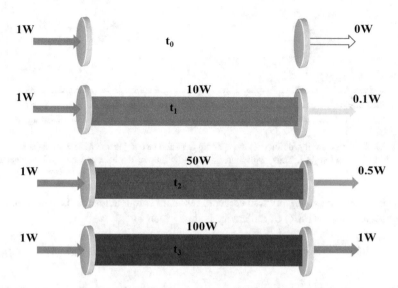

Fig. 6.13 Schematic illustration of transmission of resonant light through a cavity. At time $t = t_0$ just at the moment of shining the cavity with a beam of light of power 1 W, the intracavity photon flux is zero and nothing emerges out of the cavity (top trace). The tiny fraction that gets inside the cavity oscillates back and forth as there is a resonance. The input being continuous, the intracavity flux gradually builds up with increasing time, and in turn, the output too rises. This situation is represented in the next two traces captured at times t_1 and t_2. Eventually an equilibrium condition is reached at time t_3 as the intracavity power attains a level of 100 W when the cavity output equals its input (bottom trace)

wavelength, say λ, would pass through the cavity would depend on the length of the cavity. If the length l is such that the resonance equation, viz., $\lambda/2 \times n = l$, is satisfied, it would pass through. If not, then as the input is tuned, the immediately next wavelength that satisfies this condition would pass giving rise to one of the observed peaks of Fig. 6.9b, and as the input is continuously tuned, the cavity discretely transmits certain wavelengths.

Let us now examine the spacing between any two adjacent discrete wavelengths, viz., λ_1 and λ_2 ($\lambda_2 > \lambda_1$), that are resonant to this cavity and therefore satisfy the following condition:

$$n\left(\frac{\lambda_1}{2}\right) = (n-1)\frac{\lambda_2}{2} = l \tag{6.3}$$

In terms of frequency, the same can be rewritten as

$$n\left(\frac{c}{2\nu_1}\right) = (n-1)\frac{c}{2\nu_2} = l \tag{6.4}$$

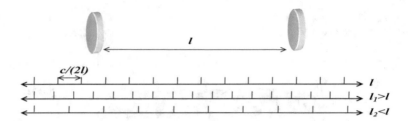

Fig. 6.14 FSR of a cavity, defined as the spacing between two adjacent cavity modes, is inversely proportional to its length l. Thus, with increasing cavity length, the FSR reduces and vice versa

The spacing between any two adjacent resonant frequencies that is also called the free spectral range (FSR) of the cavity is thus given by

$$FSR = \Delta\nu = \nu_1 - \nu_2 = \frac{nc}{2l} - (n-1)\frac{c}{2l} = \frac{c}{2l} \tag{6.5}$$

It is obvious that a cavity can support an infinite number of resonant frequencies, also called cavity modes, as depicted in Fig. 6.14. As seen here, the cavity modes can be pushed away from each other or brought closer by reducing or increasing the length of the cavity, respectively. As we shall see later in Chap. 8, this fact can be exploited to enhance the monochromaticity of laser light. It is, however, imperative that we acquire, at this point, qualitative knowledge on another aspect of the cavity, viz., its linewidth, that also has great bearing in the operation of a laser.

6.3.1   Cavity Linewidth

Heisenberg's uncertainty principle states that there is an inherent limit imposed by the nature on the combined accuracy with which a pair of certain canonically conjugate variables such as position and momentum or energy and time can be determined. If effort is expended to measure one variable more accurately at one instant, the predictability of the other variable reduces correspondingly at that instant. The broadening of spectral lines owes its origin to this. Energy and time being two such variables, an energy level of longer life will have a shorter spread in energy compared to the one that has a shorter lifetime. This spread in energy, as we now know, is manifested in the broadening of spectral width in the case of atomic or molecular transitions. This very principle, as we have seen earlier while studying "behavior of light" (Chap. 2, Sect. 2.6), also causes a beam of light to spread spatially while passing through an aperture. The ground energy level has no spread in energy as a species can remain here for an infinite time. Similar to the finite lifetime associated with an excited energy state, a practical cavity also holds light in it for a finite time. Obviously, therefore, a cavity too has a linewidth associated with it. The longer the light can reside inside the cavity, the smaller its linewidth will be,

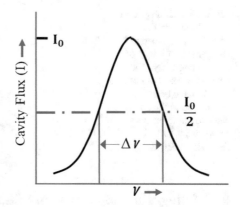

Fig. 6.15 The intracavity light flux of intensity (I) as a function of the frequency (ν) for a resonant cavity is schematically illustrated here. The cavity linewidth ($\Delta\nu$) is normally defined as the full width at half maximum (FWHM) of the resonance peak (I_0)

and the shorter the cavity lifetime is, the greater the linewidth will be. Ideally if a cavity can be constructed with two *100%* reflective mirrors placed exactly parallel to each other, light at any of its resonant frequencies can reside inside indefinitely, and consequently the cavity linewidth here will be zero. It is worthwhile at this point to pick up the intensity profile of one of the cavity resonances recorded in Fig. 6.9b, and the same is captured in Fig. 6.15. As in the case of spectral width, the cavity linewidth $\Delta\nu$ is also described as the full width of the peak at half of its maximum value. Undoubtedly, a cavity that can hold light for a longer time will be less lossy compared to the one wherein light decay faster. Obviously, therefore, a cavity with a longer length and higher reflectivity of its end mirrors that can hold light for a longer duration will be less lossy. The loss of a cavity is often quantified by a parameter called the quality factor Q of the cavity and is a measure of its power to hold light. It can be mathematically expressed as

$$Q = \frac{2\pi\nu l}{c\left(1 - R^2\right)} \tag{6.6}$$

where ν is the cavity resonance frequency, l is the cavity length, and R is the combined reflectivity of its two mirrors. Clearly, the quality factor of a cavity increases with increasing both its length and reflectivity. A cavity with a high Q value is, therefore, less lossy and consequently will have less linewidth and vice versa as schematically illustrated in Fig. 6.16. The cavity Q thus renders a measure of the linewidth $\Delta\nu$ of its resonance peaks, and the two are connected mathematically as

$$Q = \frac{\nu}{2\Delta\nu} \tag{6.7}$$

The Q value of the laser cavity, as we shall see later in Chap. 9, has a strong bearing in particular in the operation of a pulsed laser.

Fig. 6.16 A high Q cavity
is less lossy and thus will
have a smaller linewidth
compared to a low Q one
that is more lossy. The
resonance peaks of both the
cavities have been
normalized to the same
value to allow ready
comparisons of their
respective linewidths

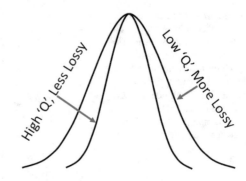

6.4   Integration of Cavity and Stimulated Emission

We now know that a cavity has the ability to transmit entire incident light, which
resonates with one of its numerous discrete modes, to the other side. The tiny
fraction of the incident light, which begins to trickle in through the high reflective
mirror, will oscillate back and forth inside the cavity and continue to grow. Soon an
equilibrium condition will be reached when the intracavity power grows to a level to
allow emergence of as much light through the cavity as is incident upon it on the
other side. At this point, the readers will be intrigued by the thought – what happens
if the source of light, instead of shining the cavity from outside, is placed inside? No
matter however much feeble the source is, light emitted at a resonant wavelength will
oscillate back and forth and continue to grow until an equilibrium is established. The
partially transmitting mirror will allow a fraction of this accumulated light to emerge
through as before. A population inverted medium itself fits the bill as an intracavity
source of light because it has the ability to generate seed photons through spontaneous emissions. Needless to say, the cavity, upon such an integration with the
amplified stimulated emission, sets up the working of a laser that we know today.
The spread in the values of energy of the two energy levels, between which the
population inversion is created, inevitably results in a corresponding spread in the
frequency of the spontaneously emitted photons. Since the stimulated emission is
seeded by these spontaneous photons, one would therefore expect a similar frequency spread in the emission of the laser too. If it were true, then the color of the
laser light would be far from pure. Fortunately, the cavity rescues the laser again and
saves its light from this ignominy. True the stimulated emission is seeded by the
spontaneously emitted photons, but the cavity, by virtue of its discretionary power,
would allow only the photons, resonant to the cavity modes, to grow through
stimulated emission. Although the cavity linewidth will restrict the spectral purity
of the oscillating mode(s), the spectral narrowing effect, described earlier in Chap. 4
and is omnipresent in a light amplifier, will greatly enrich the purity of the emitted

Fig. 6.17 (a) Owing to the finite residence time of an atom in an energy state other than ground, there is a broadening of the energy levels E_1 and E_2 in frequency as a consequence of Heisenberg's uncertainty principle (b) Thus, the population inversion or gain created between E_2 and E_1 is also broadened in frequency with its peak value occurring at the central frequency ν_0 and dropping off on either side

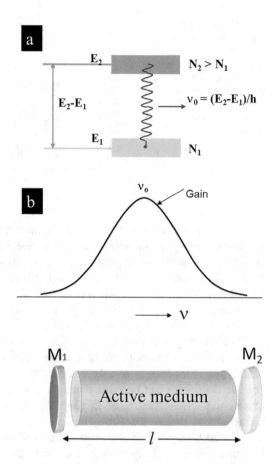

Fig. 6.18 An active medium with frequency distribution of gain as shown in Fig. 6.17b is placed inside a cavity

light. That the manifestation of the spectral narrowing in the case of an active cavity[5] can result in the shrinkage of the passive cavity linewidth to an extraordinary level will be discussed later in Sec 6.6 of this chapter.

It is pertinent here to consider a medium where population inversion has been created and is being maintained between two energy levels E_2 and E_1, which owing to their finite lifetimes are shown broadened in frequency in Fig. 6.17a. The resulting gain thus does not occur on a single frequency but is spread over numerous frequencies that, in a way, is a combination of the frequency spread of both energy levels. The highest value of gain occurs at the mean frequency viz., $(E_2-E_1)/h$, and drops off on either side exponentially (Fig. 6.17b). This active medium is now placed inside a cavity (Fig. 6.18). Let us assume a typical cavity of length *1 m* giving rise to a corresponding FSR of *150 MHz*. If the spread of gain in Fig. 6.17b be over *1000 MHz*, then just about *seven* cavity modes would reside inside the gain curve

---

[5] An optical cavity that contains a population inverted medium between its parallel mirrors is termed as an active cavity in the common parlance.

Fig. 6.19 In this case, there
are only three modes that
experience net gain, and
hence lasing occurs only on
these three modes. Mode
with higher gain also yields
higher power. The lasing is
interspersed with wide dark
zones

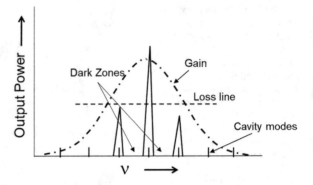

Fig. 6.19 In this case, there are only three modes that experience net gain, and hence lasing occurs only on these three modes. Mode with higher gain also yields higher power. The lasing is interspersed with wide dark zones

as shown in Fig. 6.19. We also know that although stimulated emission is seeded by the spontaneous photons emitted across the entire frequency spread over which the gain exists, it will only be sustained only on those frequencies which are cavity modes. All seven cavity modes shown in this figure possess some gain, low or high. The obvious question that arises here is "will the occurrence of stimulated emission on all these modes result in laser output on all of them?" The answer is an emphatic *no*. Normally, reflectivity of one of the mirrors of a typical laser cavity is maintained at 100% meaning no photon flux can escape through this, while a small transmission loss is provided to the other mirror through which the laser beam is coupled out as a useful loss. This loss is usually independent of the frequency over which gain persists and hence can be represented as a straight line parallel to the frequency axis usually referred to as a loss line and is also shown in Fig. 6.19. Lasing can occur on only those cavity modes for which there exists a net gain meaning whose gain exceeds the loss. The frequency spread between the two points of intersection of the loss line with the gain curve is thus the maximum possible emission width of the laser. In this example, there are only three modes for which the gain exceeds the loss, and hence lasing occurs only on these three modes (Fig. 6.19); as seen, the mode with higher gain also yields higher power. Thus, this basically represents a case of multimode laser for which the emission width is nearly *300 MHz* much purer in color compared to a typical conventional source of light. It, however, is apparent from this figure that lasing on the three cavity mode frequencies is interspaced with wide dark zones. By forcing lasing on only one cavity mode, the purity of the laser color, or in other words its monochromaticity, can be greatly improved.

6.5 Single Mode Lasing

Forcing the operation of the laser on a single mode can be accomplished if only one mode is made to satisfy the gain above loss condition. A look at Fig. 6.19 immediately suggests that it can be achieved in one of the following *three* ways as depicted in Fig. 6.20:

Fig. 6.20 Lasing on single
mode can be achieved if the
length of intercept of the
loss line with the gain curve
is smaller than the FSR. The
same can be achieved by
reducing the broadening of
the gain (top trace), by
pushing up the loss line
(middle trace), or by
pushing the cavity modes
apart (bottom trace)

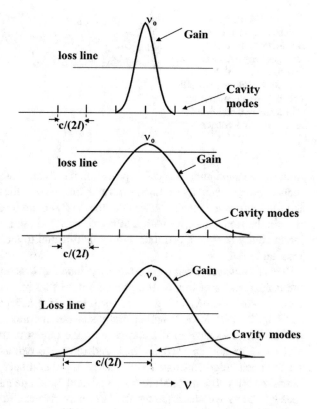

1. By reducing the broadening of the gain, the length of the intercept of the loss line
   with the gain curve can be made lesser than the FSR of the cavity. This results in
   only one cavity mode to satisfy the gain above loss condition.
2. Pushing the loss line up also makes the length of the intercept of the loss line with
   the gain curve to fall below the FSR of the cavity. This will also allow only a
   single mode to satisfy the lasing condition.
3. Reduction of the cavity length pushes the cavity modes away from each other
   allowing once again the FSR to exceed the length of intercept of the loss line with
   the gain curve. The end result will thus be a single mode laser.

Effecting a single mode in all the above three techniques, however, comes at a
price, viz., at the expense of the laser performance. For example, for a gas laser,
broadening of the gain can be reduced by reducing the operating pressure.[6] This in
turn reduces the density of the active species, thereby reducing the maximum power
extractable from the system. In the third technique, the reduction in cavity length
may have to be at the expense of the length of the active medium, thus once again

---

[6]Reduction in pressure reduces the rate of collisions that, in turn, increases the time of residence of a
species in the excited state, thereby reducing the broadening of the energy levels in accordance with
the uncertainty principle.

Fig. 6.21 A dispersive element splits the cavity modes spatially, thereby allowing only one mode at a time to receive feedback and survive. All other modes fall obliquely on the output mirror and will thus cease to receive cavity feedback

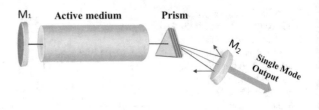

limiting the maximum extractable power. In the second case, that can be achieved by reducing the reflectivity of the mirrors and thus making the cavity more lossy which, in turn, reduces the quality factor Q of the cavity. Even if the extractable power may remain the same, a cavity with a lower Q, as we know from Sect. 6.3.1, possesses larger cavity linewidth and thus will be detrimental to the monochromaticity of the laser emission.

This disadvantage can be overcome by imposing loss selectively on all but one of the lasing modes of Fig. 6.19 by making use of an intracavity dispersive element, e.g., a prism, as shown in Fig. 6.21. The prism being a dispersive element (described in Chap. 2, Sect. 2.3.2.2) will spatially separate the modes, thereby allowing only one of them at a time to fall normally on the output mirror. Therefore, this mode alone will be able to oscillate back and forth and extract energy from the active medium resulting in a single mode lasing. All other modes that fall obliquely on the output mirror will not be able to lase at all, and the single mode emission in this case, ideally, occurs without affecting the laser performance. Imparting appropriate rotation to either the prism or the output mirror M_2 will bring another mode into lasing and thus provides tunability to the emission.

6.6   Spectral Narrowing Effect and the Purity of Laser Light

We now know that whether a laser will emit on multiple modes or on a single mode will be basically governed by three factors, viz., broadening of the gain in frequency, the free spectral range of the cavity, and its quality factor, i.e., the Q value. We also know that a cavity has a characteristic linewidth associated with it that, in the absence of any gain medium, is aptly termed the passive cavity linewidth. The frequency corresponding to each of the cavity modes will therefore have this linewidth under passive conditions. However, once an active medium is introduced inside the cavity and light begins to grow, the spectral narrowing effect manifests in a remarkable manner causing dramatic shrinkage of the passive cavity linewidth. This, in turn, imparts an extraordinary purity to the color of light emitted by the laser on each of its oscillating modes. It is appropriate here to gain insight into the linewidth-spectral narrowing connection and into the factors governing it.

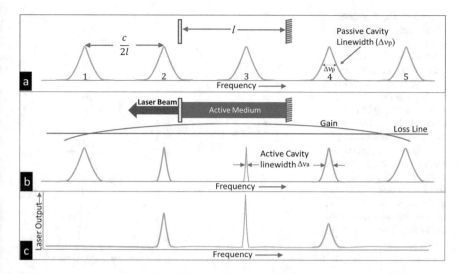

Fig. 6.22 Manifestation of spectral narrowing effect in the operation of a laser. a) A typical cavity and five adjacent modes and their broadening under passive condition are shown. b) With the onset of lasing, upon introduction of an active medium inside the cavity, there is a dramatic reduction of the cavity linewidths as a result of spectral narrowing. The spectral width is least for the mode that experiences the highest gain. The modes that do not experience any net gain undergo no narrowing. c) The spectral content of the lasing modes is also translated into the spectrum of the laser emission. The color is undoubtedly purest for the mode that sees the highest gain. The mode with the highest gain also produces highest power

Let us begin by considering a typical cavity that comprises one total and another partial reflector separated by a distance l as depicted in Fig. 6.22a. Five adjacent cavity modes along with their frequency spread are also shown. As the cavity is initially passive, the frequency broadening of all its modes equals the natural passive cavity linewidth ($\Delta \nu_p$). As the process of lasing begins with the introduction of an active medium inside the cavity, the spectral narrowing effect also sets in (Fig. 6.22b). This causes dramatic shrinkage of the passive cavity linewidth, greatly enhancing the spectral purity of the laser light. It is appropriate at this point to examine as to how the introduction of the gain medium alters the passive linewidths of each of the five cavity modes captured in trace (a) of Fig. 6.22. We know that the spacing in frequency between two successive cavity modes, the FSR of a cavity, is given by $c/2l$ where l is the cavity length and c is the velocity of light. Let us assume that the broadening of the gain is large enough not only to encompass all five modes but extend even far beyond. The situation is schematically represented in trace (b) of this figure. The cavity, as we now know, also has an inherent loss associated with it determined by its quality factor Q. The cavity Q is governed by the length and the combined mirror reflectivity and can be represented, as we know, as a straight line parallel to the frequency axis. As we can see, lasing is possible only on three modes, viz., mode numbers *2*, *3*, and *4* for which gain exceeds loss. Clearly no amplification of light will be possible for modes *1* and *5* as their respective gain lies below the loss

Fig. 6.23 The frequency distribution of spontaneously emitted photons, permitted to oscillate by the cavity on a particular mode, is shown in the upper trace. Upon travelling through the gain medium, the photons at the mode center will multiply more rapidly compared to that at the wings of the distribution. This will inevitably cause a narrowing of the cavity linewidth and, in turn, the emission width of the laser (lower trace)

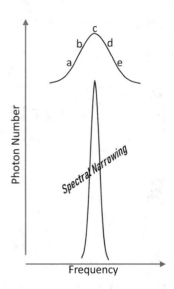

line. Let us first focus our attention on mode *3*, which possesses the maximum gain, and examine as to how the manifestation of the spectral narrowing effect impacts its passive linewidth. To simplify the understanding, we label five different frequency points *a*, *b*, *c*, *d*, and *e* across this passive linewidth as shown in the upper trace of Fig. 6.23. Obviously, the number of photons arising out of spontaneous emission will be maximum at the central frequency of the mode, indicated by the point *c*. Their number will continue to drop as we gradually move away from the center on either side as indicated by points *b* and *d* as well as points *a* and *e*. As these spontaneously emitted photons with such a distribution in frequency begin to travel through the population inverted medium, the photon's number at the central frequency will therefore multiply through stimulated emission at a faster rate compared to that at the wings. This means that the number of photons at *c* will build up more rapidly than that at *b* or *d*. Likewise the number of photons at *b* or *d* will multiply more rapidly compared to that at *a* or *e*. This is a positive regenerative effect and will cause extreme spectral narrowing of the cavity linewidth (lower trace). Needless to say, the extraordinary purity of the color of the laser light owes its origin to the narrowing of the linewidth of the laser cavity. The higher the gain of the medium is, the more pronounced this effect will be; hence, the spectral width of mode number *3* will experience the maximum contraction due to lasing. Of the three lasing modes, mode number *4* has the lowest gain and, therefore, will experience the least effect of narrowing. The spectral narrowing of mode number *2* will lie between *3* and *4*. Modes *1* and *5* do not show any narrowing effect as they do not get to lase in the first place. The distribution of the output power of this multimode laser on these three oscillating modes is schematically shown in trace *c* of Fig. 6.22. The strong regenerative component intrinsic to the phenomena of cavity linewidth narrowing can, in principle, result in a mind-boggling purity[7] of the laser light even for a laser of moderate power. In reality, however, the light emitted by a practical laser is much less pure for the following reason:

We know from Sect. 6.3 that the resonant condition for n^{th} cavity mode of wavelength λ for a cavity of length l is

$$n\frac{\lambda}{2} = l \tag{6.8}$$

Adding or subtracting $\lambda/2$ on both sides of this equation, we obtain

$$(n \pm 1)\frac{\lambda}{2} = l \pm \frac{\lambda}{2} \tag{6.9}$$

This means that if λ is the wavelength of the nth resonant mode of a cavity of length l, then it becomes the $(n + 1)^{th}$ resonant mode for a cavity of length $(l + \lambda/2)$ or the $(n - 1)^{th}$ resonant mode in the case of a cavity of length $(l - \lambda/2)$. This is equivalent to saying that if a laser, of cavity length l, is operating at wavelength λ and if its cavity exhibits a length change of $\lambda/2$, then the oscillating mode frequency will sweep through the entire FSR of $c/2l$. To appreciate the detrimental effect of this fact on the ultimate achievable purity of the laser light, let us consider the representative example of a common laser such as the helium-neon laser. A *He-Ne* laser can operate on a visible (red) wavelength of *0.6328* micrometer, and assuming a *1-m-long* cavity, its *FSR* works out to be *150 MHz*. If the cavity length of the laser fluctuates by $\sim \pm 0.03$ µm, i.e., $\sim \pm \lambda/20$, the laser mode(s) will then unmistakably exhibit a frequency jitter of ~30 MHz. Such tiny fluctuations in the cavity length are inevitable and often caused by ground borne vibration/mechanical disturbances, and airborne sound waves, that can readily reach the mirrors forming the cavity. The variation in the refractive index of the lasing medium during the process of lasing is another factor that can greatly contribute to the cavity length fluctuation[8] in the pulsed operation of certain lasers. The *30 MHz* emission linewidth of the red He-Ne laser is equivalent to a spread of $0.4 \times 10^{-3}\,A^0$ around its central wavelength of $6832\,A^0$. In contrast, a low-pressure sodium vapor lamp has a much wider spread $(0.1\,A^0)$ around its central wavelength $5890\,A^0$. The *He-Ne* laser light is thus at least *250* times purer than the sodium light that was regarded as vastly monochromatic before the advent of lasers! As we shall see in a latter chapter (Chap. 8), it is possible to immensely improve the purity of the laser color by holding its cavity mirrors with

---

[7]Chandra Kumar Patel, who invented CO_2 lasers in 1963 and served as an executive director of AT&T Bell Laboratories, Murray Hill, NJ, presented a theoretical estimate of laser linewidth to be less than even a hertz in his well-read review paper on gas lasers [22]. It may be of interest to note here that CO_2 laser finds wide applications in industry, medicine, military, skin resurfacing, and many more.

[8]The cavity length is basically the product of the geometrical separation l of the mirrors and the r.i. of the medium enclosed between them. For a gas laser, the r.i. of the medium is ~ 1, and the cavity length can therefore be approximated as l.

nearly vibration-free structure and taking measures to actively lock the lasing mode to a preassigned frequency. Lasers with such improved frequency stability are prerequisite for spectroscopic or long-distance interferometric applications.

6.7   On the Stability of a Cavity

Having come this far, we have got a glimpse on the pivotal role that a cavity plays in enhancing both the magnitude and quality of light manufactured by the laser. The preceding section has touched upon on the physics, albeit qualitatively, behind the cavity assisted enhancement of the purity of the laser light. We also know that by virtue of the property of stimulated emission, the cavity has an innate ability to produce a directed beam of light from an active medium. The obvious question that arises here is "will this beam of light emitted by the laser propagate like an ideal parallel beam of light in the free space?" The answer is an emphatic no. In this context, it would be worthwhile to take another look at the phenomenon of diffraction of light described in Chap. 2.

A parallel beam of light, upon passing through a tiny aperture, of size comparable to its wavelength, exhibits an angular spread $\approx \theta$, the minimum value of which is given by

$$\theta \approx \frac{\lambda}{d} \tag{6.10}$$

where λ is the wavelength of light and d is the diameter of the aperture, assuming it to be a circle. This is equivalent to saying that having passed through an aperture, the parallel beam loses its parallelism and diverges upon propagation through space and is termed as "diffraction limited divergence." If the size of the aperture becomes much too big compared to the wavelength, then it is apparent from the above equation that θ tends to zero. In the limiting case of an aperture of infinite dimension, the beam does not diverge at all meaning that the parallel beam passes through an infinite aperture unhindered, almost oblivious to its presence. If instead of a parallel beam we have a diverging beam to begin with, then the aperture induced diffraction will result to an enhancement of its geometrical divergence. All these three situations are schematically represented in the traces of Fig. 6.24.

We are now in a position to answer the question on the directionality of laser light posed at the beginning of this section. In the case of a cavity formed by two plane parallel mirrors, the spontaneously emitted photons that move parallel to the cavity axis will grow in numbers through the process of stimulation as they oscillate repeatedly back and forth between the mirrors. However, the finite cross section of the gain medium makes the situation far removed from what is depicted in Fig. 6.24b and will inflict diffraction limited divergence to the amplified parallel beam of light that is confined within it. The aperture effect imposed by the finite transverse dimension of the system is in a way akin to the case represented in the trace of

Fig. 6.24 The phenomenon of diffraction always causes a parallel beam of light to diverge as it passes through a narrow aperture of dimension comparable to its wavelength. The angle of divergence (θ) is called the diffraction limited divergence (**a**). The beam, however, passes through a much wider aperture unhindered (**b**). In case of a diverging beam passing through a narrow aperture, diffraction renders its actual divergence (blue) more than the geometrical divergence (red) (**c**)

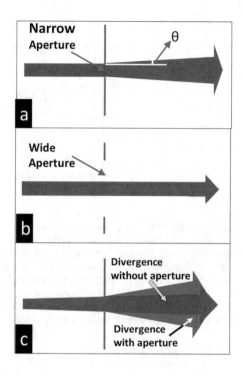

Fig. 6.24a. The minimum divergence with which the laser beam emerges through a parallel mirror-based cavity is therefore limited by the effect of this natural diffraction that is inevitable. However, in reality, the divergence of a conventional laser will exceed this value for the reason qualitatively depicted in the traces of Fig. 6.25. An optical cavity comprising of two plane mirrors, one partial and the other fully reflective, is shown in the top and middle traces of this figure to underline the extreme sensitivity of a plane parallel cavity to the misalignment of its mirrors. When the mirrors are perfectly aligned, the spontaneous photons emitted from the active medium, along the axis of the cavity, will oscillate back and forth and grow through stimulated emissions. This, as we know now, will result in the formation of a beam of light inside the cavity, a fraction of which will emerge as a parallel beam of laser light ignoring the effect of natural divergence (top trace). On the other hand, for a laser with reasonable length of its cavity, even an insignificant misalignment will make the light bouncing back and forth between its two mirrors quickly fly away (middle trace). This will not allow any significant amplification of light through stimulated emission inside the cavity let alone giving out a laser output even of any modest power. Such loss of light will obviously reduce with the reduction of cavity length and can be practically avoided for extremely small lasers such as microchip or semiconductor lasers. For more conventional lasers, the extreme sensitivity to misalignment is overcome by replacing at least one of the plane mirrors of the cavity by one with a curved geometry. The focusing power of the curved mirror allows trapping of light even if it is not precisely parallel to the axis of the cavity as is

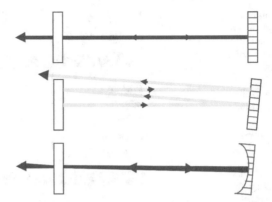

Fig. 6.25 When the two plane mirrors forming the cavity are exactly parallel to each other, back and forth oscillations of the spontaneously emitted photons are possible allowing the operation of a laser (top trace). However, even an insignificant misalignment of the mirrors will make light reflected from one mirror to quickly miss the other, disrupting the operation of the laser (middle trace). This disadvantage can be overcome by replacing at least one of the plane mirrors with a curved one (bottom trace). The converging power of the curved surface does the trick by directing the off-axis light back into the cavity. As can be seen, the stability of the cavity comes here at the expense of the directionality of the laser emission

Fig. 6.26 The cavity length l and the radius of curvature R_1 and R_2 of the mirrors are the parameters that govern the cavity stability

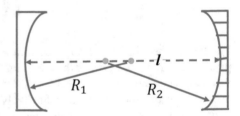

evident from the bottom trace of this figure. The usage of a curved mirror therefore adds stability to the cavity by making it less sensitive to the mirror misalignment. The stability, however, comes at the expense of the directionality of the laser beam. The ability of the curved mirror to converge light inside the cavity understandably adds to the divergence of the emerging laser beam.

The cavity stability, therefore, is critically dependent on the geometry and radius of curvature of the mirrors and their separation. The stability criterion of a cavity can be mathematically expressed as

$$0 \le \left(1 - \frac{l}{R_1}\right)\left(1 - \frac{l}{R_2}\right) \le 1 \qquad (6.11)$$

where R_1 and R_2 are the radii of curvatures of the two mirrors and l is their separation and are depicted in Fig. 6.26. In the case of a plane-plane cavity, $R_1 = R_2 = \infty$, and the product $\left(1 - \frac{1}{R_1}\right)\left(1 - \frac{1}{R_2}\right)$ equaling exactly 1 lies precisely on the stability boundary defined by the above condition. A plane-plane cavity can therefore be considered to be only theoretically stable, and it is no wonder that such a cavity is extremely sensitive to misalignment. The above stability criteria clearly establish that the mirror's curvatures and their separation alone govern the cavity stability and other parameters such as laser gain, and wavelength has no bearing on it. The length of the active medium basically limits the minimum value of l, and there can be numerous combinations of R_1 and R_2 that would make the cavity stable. As we would see in the following sections, the intended application of the laser and its type will be the primary criteria for selecting the appropriate combination of cavity parameters.

6.8   Geometry of the Reflective Surface, Intracavity Wave Front, and Gaussian Beam

It is pertinent here to focus on the wave front of the laser beam that establishes itself by bouncing back and forth very many times inside a stable cavity. Toward simplifying the interpretation, we disregard here the presence of the aperture effect. Let us first consider the case of a plane-plane cavity. Obviously for a light beam to survive inside a resonator, its wave front must match exactly with the curvature of the cavity mirrors. As the reflective surface of the plane mirror is planar, a parallel beam of light whose wave front is planar can therefore only survive in a plane-plane cavity. Such a cavity produces a laser beam with usually minimum divergence. However, as discussed before, it suffers from extreme sensitivity to mirror misalignment. A cavity comprising a plane and curved or both curved mirrors, with inbuilt stability, is therefore a more popular choice for the construction of a laser resonator. Discernably, the curved reflecting surface(s) of such a cavity will preclude the existence of a parallel beam of light inside it unlike in the plane-plane mirror case. The obvious question that arises here is "what will be the nature of the intracavity laser beam with wave fronts fitting into the curved and plane reflecting surfaces or both curved surfaces?" The answer is a Gaussian beam that fits the bill here. While we choose to bypass the intricate mathematics, a small digression to acquire a qualitative understanding of the nature of a Gaussian beam that is practically entwined with the operation of a laser will be helpful. A typical cross-sectional view of the Gaussian beam is schematically illustrated in Fig. 6.27 for a stable cavity formed with two concave mirrors having the same radius of curvature. A symmetrical cavity is being considered here for ease of understanding. We shall of course subsequently consider more general cavity configurations to consolidate our understanding of this beam. Two inverted curved lines extending from one cavity mirror to the other, as seen in this figure, basically represent the Gaussian beam. The beam actually has no sharp edge, and its two longitudinal boundaries on either side are defined by the locus of the respective points where the intensity I has dropped down to $1/e^2$ of its corresponding value I_0 on the central line.

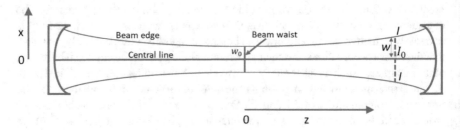

Fig. 6.27 The schematic of a Gaussian beam for a stable cavity formed with two concave mirrors of identical radius of curvature

The Gaussian beam is characterized by two fundamental parameters, viz., the beam radius (w) and the radius of curvature (R) of the beam's spherical wave front.[9] The beam radius is defined as the transverse distance between the beam boundary and the beam center. It has a minimum value w_0 called the beam waist that understandably occurs midway between the cavity mirrors in this example. The radius of the beam increases on either side of the beam waist, and the two are mathematically related through

$$w(z) = w_0 \left[1 + \left(\frac{\lambda z}{\pi w_0^2} \right)^2 \right]^{\frac{1}{2}} \tag{6.12}$$

where $w(z)$ is the beam radius at a distance z from the beam waist w_0 and λ is the wavelength of the laser. The intensity profile of a Gaussian beam across a beam radius w as a function of the distance x from the center is given by

$$I = I_0 e^{-2x^2/w^2} \tag{6.13}$$

and is schematically illustrated in Fig. 6.28. The radius of curvature (ROC) of the wave front of the Gaussian beam as a function of z, i.e., the distance from the beam waist, is given by

$$R(z) = z \left[1 + \left(\frac{\pi w_0^2}{\lambda z} \right)^2 \right] \tag{6.14}$$

A thorough examination of this equation reveals that the ROC of the wave front at the beam waist is infinite, sharply drops on either side, and then begins to rise again as we move further away on both sides. Clearly for a cavity to be stable, the wave

---

[9] Wave front is the locus of all the points where waves starting simultaneously from the source arrive at a given instant of time. The wave front of a point source will therefore be a sphere, and the rays of light will be normal to the surface of the sphere.

Fig. 6.28 The intensity profile of a Gaussian beam. The boundary of the beam is defined as the point where the intensity drops down to $1/e^2$ of its peak value and w is the corresponding beam radius

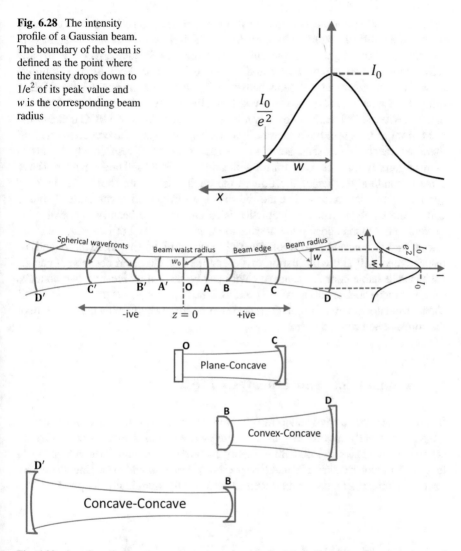

Fig. 6.29 As a Gaussian beam propagates through space, both its radius and radius of curvature of its wave front change. A stable resonator cavity is possible with plane-concave, convex-concave, and concave-concave configurations whenever the mirror curvatures match the corresponding spherical surfaces of the Gaussian beam wave front. The Gaussian beam's intensity profile recorded in the preceding figure has been replicated here corresponding to a beam radius of w for a better perception

front must fit exactly into the reflective surface of the two end mirrors forming a cavity. The values of Gaussian beam radius and its radius of curvature as a function of z evaluated from Eqs. 6.12 and 6.14 for a typical beam waist parameter and lasing wavelength, respectively, are graphically represented in Fig. 6.29. As the cavity is symmetric, the two sides of the beam waist are just mirror images of one another. As seen, radius of the Gaussian beam increases monotonically on either side of the beam waist quite unlike the ROC of its wave front, which shows a minimal behavior with increasing z. This figure also presents a guideline, albeit qualitatively, as to how ROC of the mirrors is to be chosen and located to form a stable cavity. For example, if we have to construct a planoconcave cavity, the plane mirror, whose ROC is infinity, must be located at the beam waist, and the concavity of the other mirror will of course decide its location. For instance, if the concave mirror's ROC matches the wave front at C, it is to be located right there. The plane and the curved mirror will therefore match the Gaussian beam's wave front at O and C, respectively, as shown in this figure. Another concave mirror with a different ROC will understandably have a correspondingly different location in the cavity. In reality, however, the only surviving Gaussian mode will be the one with wave front radii of curvature matching that of the cavity mirrors. The practice is to engineer the beam by choosing the mirrors rather than choosing the mirrors to match the beam. For instance, a convex mirror and a concave mirror of the right curvature when placed at B and D, respectively, will allow oscillation of the segment of the beam between B and D. Likewise, a stable concave-concave cavity can be formed by locating two concave mirrors of appropriate curvature at D' and B. Needless to say, a countless number of stable resonator cavity configurations are possible simply by varying the curvature of the concave and convex mirrors.

6.9   Longitudinal and Transverse Modes

It is now clear that a cavity supports only those wavelengths half of which times an integer fits exactly into its length. Two such wavelengths λ_1 and λ_2 ($\lambda_2 > \lambda_1$) are schematically shown in Fig. 6.30 to satisfy such a resonant condition in a cavity of length l for integer values of 9 and 8 respectively. For easy understanding, the cavity length in this figure has been assumed to be only marginally longer than the

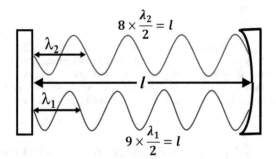

Fig. 6.30 Simultaneous oscillation of two adjacent longitudinal modes is schematically illustrated. Cavity resonance condition is shown to be satisfied for integer values of 8 and 9 for the two wavelengths. In reality, the wavelength is far too small compared to the cavity length, and the integer value will be much too high

wavelengths. In reality, however, the integer value can be astronomically high. For instance, for a *1*-m-long visible laser, the integer value will run into several millions. Sticking for now to the situation, primarily representational, described in Fig. 6.30, we realize that light at these two resonant wavelengths travels parallel to the axis of the cavity. These resonant modes that are axial to the cavity are called longitudinal modes. Radiation at wavelengths λ_1 and λ_2 therefore corresponds to the *eighth* and *ninth* longitudinal modes of this cavity, respectively. From the stability point of view of the cavity, as we know now, the spatial intensity profile of both these modes must be Gaussian in nature that in the common parlance is denoted as the TEM_{00} mode.[10] (The general nomenclature of an oscillating lasing mode is elaborated later in this section.) Thus, λ_1 and λ_2 are the *eighth* and *ninth* longitudinal TEM_{00} modes of the cavity, respectively, and are denoted as TEM_{008} and TEM_{009} modes. Equivalently we can say that TEM_{00q} is the *q*th longitudinal TEM_{00} or Gaussian mode of the cavity. The longitudinal mode index q can vary, as we know, from 1 to $\propto$ meaning that a cavity has an infinite number of resonant modes, and the frequency spacing between any two adjacent mods is $c/2\,l$. These Gaussian longitudinal modes in a real laser system will involve as many radiation beams traveling exactly the same path through the active medium. Of course, only those modes for which gain exceeds the cavity loss will eventually grow.

So long we have restricted ourselves only to the rays axial to the cavity, but in a real laser system, photons originating from the spontaneous emissions may also travel obliquely to the cavity axis. Will these off-axis rays have any bearing on the operation of a laser? They understandably will have no influence on the operation of a laser based on the plane-plane cavity that is extremely sensitive to misalignment, and the oblique photon reflected from one mirror may therefore easily miss the other. However, what happens in the case of a curved mirror-based cavity that has the capability to withstand misalignment of the cavity mirrors to a certain extent? The plot obviously thickens here and the situation warrants a deeper look. We once again consider the previous planoconcave cavity of length l but now with only one longitudinal mode of wavelength λ_1 and mode index 9, for ease of representation, and the same is schematically shown in Fig. 6.31. Consider now an off-axis ray CD to which the cavity makes an intercept of length l^+. This being marginally longer than l, the cavity resonance condition will now be satisfied here with the same mode index 9 but for a wavelength λ_2, which would be correspondingly longer than λ_1. This mode will involve slightly different optical paths through the gain medium and thus would emerge a little oblique to the TEM_{009} mode. Clearly this oblique mode will exhibit a rotational symmetry, and its counterpart would emerge through the laser along $C'D'$, as shown also in this figure.

These two oblique emissions, in conjunction, define what is known as a transverse mode. This mode will have the same index 9 as that of the TEM_{00} mode shown

---

[10]TEM stands for transverse electromagnetic, and the prefix *00* basically means that the field of the Gaussian beam decays asymptotically to zero in the transverse directions to its motion. This will be better understood when we learn about higher order transverse modes.

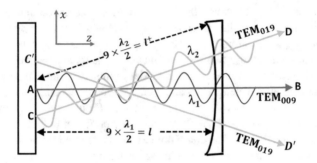

Fig. 6.31 Simultaneous oscillation of two modes with same index, viz., 9, one longitudinal and the other transverse, is schematically illustrated. The on-axis mode is designated as TEM_{009}. The two oblique beams, in conjunction, define the transverse mode that is designated in this particular case as TEM_{019}

here. However, the intensity profile of this mode will greatly differ from that of a Gaussian beam, and the same is shown in Fig. 6.32 along with the spatial impression or pattern this laser beam will make on a screen. The intensity profile of a TEM_{00} mode and its spatial pattern are also shown in this figure for comparison. An oscillating lasing mode is usually denoted as TEM_{mnq}; q is the mode index, i.e., the integer value for which the cavity resonance condition is satisfied, and m and n are the number of dips the spatial intensity profile of the lasing mode exhibits in the X and Y directions, respectively. For a Gaussian beam, both m and n are always zero. This means that for this mode, the intensity of the beam does not experience any dip and asymptotically decays to zero along both the X and Y directions. The spatial intensity profile of a Gaussian mode along the X direction, shown in the top trace of Fig. 6.32, also holds true for the Y direction. This, in turn, results in the emergence of a spatially symmetric beam, and the same has also been shown here alongside its intensity profile. Another look at Fig. 6.31 clearly points to the fact that the intensity of the transverse mode, a culmination of the oscillation on two oblique cavities, will, on the other hand, experience a dip along the cavity axis and two peaks on either side. The intensity profile of this mode and its spatial pattern are displayed in the upper middle trace of Fig. 6.32. We can therefore readily assign the m and n values to this mode as 1 and 0, respectively, meaning that this particular transverse mode can be identified as TEM_{10q}. If, on the other hand, the two oblique cavities befall along the Y axis, then m and n will, respectively, assume the values 0 and 1, and, consequently, the transverse mode will now be designated TEM_{01q}. The intensity profile and spatial pattern of this mode are recorded in the lower middle trace of this figure. It must be noted here that there is no sacrosanctity of the two oblique cavities shown in Fig. 6.31 along which the *ninth* mode exhibits resonance. As a matter of fact, there could be countless numbers of such oblique pairs of lines along which resonance conditions for this mode would be satisfied, albeit with correspondingly different wavelengths. It is not surprising that a laser is known to operate simultaneously on multiple transverse modes and each of them can be classified based on its respective m and n parameters. As a representative example, the intensity profile and

Fig. 6.32 The intensity profile of the Gaussian mode is shown in the upper trace, and the same for a few transverse modes along with their corresponding m and n parameters is shown in the remaining traces. The spatial pattern of the laser beam for all these modes is also shown alongside

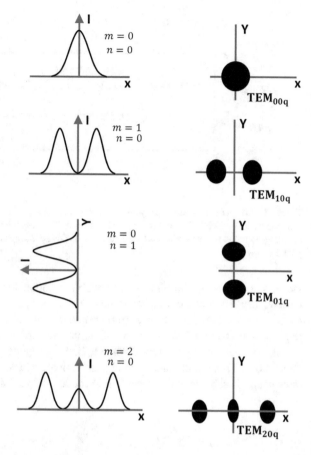

spatial pattern of the lasing beam, identified to be TEM_{20q}, are illustrated at the bottom trace of this figure. It needs to be appreciated here that Fig. 6.32 displays only a few transverse modes belonging to the family of qth Gaussian mode viz., TEM_{00q}. If the active medium supports oscillation on another Gaussian mode, say $TEM_{00(q + 1)}$ or $TEM_{00(q-1)}$, another set of transverse modes belonging to the family of this Gaussian mode may appear in the emission of the laser. If the two modes share the same volume of the active medium, then understandably, the mode that has intrinsically higher gain would compete the other out. If they utilize largely different regions of the active medium, lasing would occur simultaneously on both of them. For instance, it is readily apparent that the TEM_{00q} and TEM_{10q} modes shown in Fig. 6.32 utilize largely different volumes of the gain medium and thus would grow simultaneously. The off-axis nature of the cavity on which a transverse mode oscillates imparts an additional divergence to the corresponding emission of the laser. Higher is the order of the transverse mode, the more also will be its divergence. A laser is often required to give out a low divergent beam of light, which essentially calls for the elimination of the transverse modes. This is possible by introducing

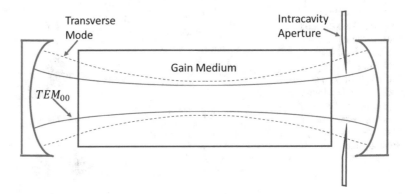

Fig. 6.33 An aperture placed appropriately between the gain medium and one of the cavity mirrors can impart loss selectively on the transverse modes. The transverse Gaussian mode, however, continues to move almost oblivious to the presence of this aperture, and the laser, in turn, emits on a pure TEM$_{00}$ mode

losses selectively on the transverse mode(s). Another look at Figs. 6.31 and 6.32 clearly suggests that the transverse modes are always larger in diameter than the Gaussian mode and survive by burning population inversion that resides away from the axis. Insertion of an aperture of appropriate diameter will therefore allow unhindered passage of the low divergent Gaussian mode while blocking the transverse mode. This arrangement, which results in the emission of a pure *TEM$_{00}$* mode laser beam, is schematically described in the illustration of Fig. 6.33.

6.10   Unstable Cavity

Fig. 6.33 of the preceding section establishes the fact that a laser, when forced to emit on the TEM$_{00}$ mode alone, invariably retains a significant amount of unutilized population inversion. The narrowness of this mode basically precludes its interaction with a sizeable fraction of population residing a little away from the lasing axis. The unused energy stored in the medium as its internal energy eventually decays out through spontaneous emission and serves no practical purpose. A TEM$_{00}$ mode in conjunction with one or more transverse modes can enhance the energy extraction efficiency of the laser. This, however, comes at the expense of the directionality of the laser emission as the transverse modes are intrinsically more diverging in nature. A novel configuration of the resonator cavity, known as an unstable cavity, can be employed to enhance the energy extraction efficiency of the laser but not compromising its spatial quality. Such a cavity often comprises of appropriately separated two suitably curved mirrors, one convex and the other concave, and the same is shown in Fig. 6.34. The curved surfaces of both mirrors are fully reflective, and hence the laser beam cannot emerge here in a conventional way through one of the mirrors. It instead exits around the edge of the convex mirror, usually smaller in

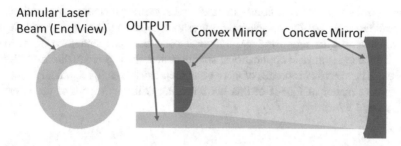

Fig. 6.34 An unstable resonator cavity comprising of a convex and a concave mirror. It is applicable to only wide aperture lasers possessing intrinsically very high gain and capable of giving out a low divergent beam like the TEM_{00} mode

size, as an annular beam. In this configuration, the initial few spontaneously emitted photons grow axially between the two mirrors and then expand toward the convex mirror. Through reflection and further expansion to the larger concave mirror, it eventually encompasses practically the entire volume of the active medium before diffracting out of the cavity as a highly energized laser beam. The beam understandably emerges in the shape of a doughnut as its central portion is obscured by the convex mirror. Although this cavity is capable of burning down the entire inversion giving out a low divergence beam, its applicability, however, is limited only to lasers possessing intrinsically very high gain. This is because in an unstable cavity, unlike in the case of a stable resonator where the beam moves inside the cavity repeatedly many times, the beam is capable of making only a few passes before exiting.

6.11 Cavity and Properties of Laser

We now know that the cavity has given a laser its two most important attributes, viz., its directionality and monochromaticity. The directionality in the emission is of course fundamental to a stable cavity laser. It is, however, inevitable that the beam given out by a laser cavity will have a divergence and the degree of divergence depends on the nature of mode, Gaussian or transverse, that the cavity supports. If the application requires a beam with lower divergence, cavity can be designed to suppress higher order modes. The laser cavity has built into it the capability to give out light that is remarkably pure in color. This is because the cavity allows oscillation only on discrete frequencies and also facilitates the process of stimulated emission driven spectral narrowing. The unique properties of a laser such as its brightness, focusability, intensity, spatial coherence, and temporal coherence are essentially manifestation of either of the above two basic attributes. Furthermore, the tunability and short pulse generation capability of a laser also arise from the manipulation of one or another parameter of the cavity. Clearly, therefore, a laser owes all its unique properties to the brilliance of the cavity it is made up of. A particular application of the laser takes advantage of its one specific property. For instance, the capability of a laser to offer high intensity comes in handy for material processing applications, its focusability and ability to generate ultrashort pulses makes it appropriate for micromachining or microsurgery, temporal coherence is mandatory for spectro-scopic or interferometric applications, spatial coherence is prerequisite for hologra-phy etc. The specific properties of a laser and their use in specific applications will be the subject matter of Part-II of this book dealing with the impact of the lasers on science and humanity.

Chapter 7
Continuous and Pulsed Lasers

7.1 Introduction

Humanity's familiarity with both continuous and pulsed sources of light dates back to time out of mind. The most glaring example of a continuous source of light is unmistakably the Sun that has been illuminating the entire solar system incessantly from time immemorial. From the humble beginning of the rudimentary camp fire, man-made continuous light sources have now graduated into vastly sophisticated ones capable of accomplishing tasks unthinkable not so long ago. That light can also be emitted intermittently was most certainly perceived for the first time ever by our great ancestors when the brilliant sparks of lightning pierced the sky during a thunderstorm. The development of a pulsed source of light that began its modest journey from the stone age with the creation of a flash of light upon rubbing a piece of stone with another has been a key to the evolution of civilization. It is no wonder, like the natural and artificial sources of light, that lasers also operate both in the continuous and pulsed modes. The first laser, a solid-state (ruby) laser, that Maiman operated in the spring of 1960 at Hughes Research Laboratory, USA, was a pulsed laser. Ruby basically served here as the host material into which the light producing Cr atoms were embedded. The next laser that was operated shortly afterward by Ali Javan (1926–2016) and his coworkers at the Massachusetts Institute of Technology, USA, was a continuously working laser. It made use of a mixture of helium and neon; neon served here as the active medium, while the presence of helium facilitated the realization of population inversion.

Continuously working lasers are usually referred to as continuous wave or *cw* lasers. They operate in the steady state regime meaning that under a steady condition of pumping, the laser yields a steady output in time. Maintaining population inversion over such a steady state condition is understandably an uphill task in a three-level energy scheme. It is not surprising that lasers suitable for exhibiting continuous operation usually work on a four-level energy scheme. Power emitted by

D. J. Biswas, *A Beginner's Guide to Lasers and Their Applications, Part 1*,
Undergraduate Lecture Notes in Physics,
https://doi.org/10.1007/978-3-031-24330-1_7

the *cw* lasers is typically not very high and does not normally exceed the kilowatt range. A pulsed laser, on the other hand, does not need to operate under steady state conditions, and consequently creation and sustenance of population inversion, over a brief period, is feasible even in a three-level scheme. Notably, the operation of the very first laser was based on a three-level energy configuration. In fact, the shorter the duration of the laser pulse is, the easier it is to meet the condition of population inversion. Today, we have lasers that are capable of producing pulses of femtosecond duration, a time practically of the order of the period of optical cycles. Such lasers are known to be invaluable in the study of transient phenomena occurring on the femtosecond timescale. The shortness of the pulse has made it possible to achieve a peak power exceeding thousands of petawatts. Incidentally, the magnitude of power from sunlight that strikes the Earth's atmosphere at any given instant lies below 200 petawatt. Regarding the application of lasers, the *cw* and pulsed lasers complement each other in a remarkable manner. This chapter aims to follow a largely nonmathematical approach to impart to the mind of the readers a comprehensive knowledge on the working of both *cw* and pulsed lasers.

7.2   Limitations of CW Lasers

It is pretty clear that a laser will continue to emit only as long as the condition of population inversion endures. Thus, for it to work continuously, the lasing medium needs to be pumped in a nonstop manner. To appreciate the consequence of depositing energy incessantly into the active medium, it is imperative to take a closer look at the energy level picture of a laser described earlier in Chap. 5. Let us first consider the operation of a laser based on a three-level energy scheme, as illustrated in Fig. 7.1. As the entire population initially resides in the ground state that also happens to be the lower lasing level, realization of population inversion essentially calls for transferring more than 50% of this population to the upper laser level. This, as we know, is effected in two steps. The act of pumping first deposits energy from an external source into the lasing medium to selectively transfer the

Fig. 7.1 Schematic of a laser based on a three-level energy configuration

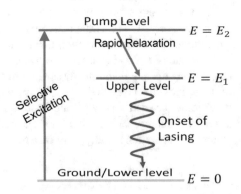

population from the ground to the pump level. The population arriving in the pump level then rapidly relaxes, usually nonradiatively,[1] into the upper laser level resulting in the realization of population inversion. The practical difficulty in transferring such a vast population from the ground to the lasing level and then maintaining the inversion in a continuous manner becomes apparent if we probe the situation more intricately. Let us assume the initial density of the ground state population to be N_0 of which at least $N_0/2$ must be transferred to the upper laser level for the onset of lasing. If the spontaneous lifetime of the upper level is τ_{sp}, then the rate of emission of photons through spontaneous decay from this level per unit volume of the active medium is essentially

$$\sim \frac{N_0}{2\tau_{sp}}$$

The energy of each of these spontaneously emitted photons is E_1, and assuming the volume of the active medium to be V, the loss of power through the spontaneous fluorescence can be readily shown to be

$$\frac{N_0 E_1 V}{2\tau_{sp}}$$

A fraction of the incident power is also dissipated nonradiatively as the population is transferred from the pump to the upper laser level. Thus, the minimum power, also called critical fluorescence power, that would be required to be deposited in a continuous manner to make up for the loss due to spontaneous decay is given by

$$\frac{N_0 E_1 V}{2\tau_{sp}} \times \frac{E_2}{E_1}$$

Plugging in the values for a typical three-level laser, e.g., a ruby laser, the critical fluorescence power can be shown to run into several kilowatts. We thus see that continuous operation of a laser based on a three-level scheme, although not forbidden, calls for an intense level of pumping.

This problem can be readily overcome by going for a four-level energy scheme, already elaborated in Chap. 5, and the same has been illustrated in Fig. 7.2. The lower laser level, now decoupled and sufficiently removed from the ground level, is essentially empty. Inversion of the population in such a system is possible with relative ease as the transfer of even a lone atom/molecule from the ground state to the

---

[1] Nonradiative relaxation is the transition from a higher to lower energy level not involving emission of any light, meaning it is a radiation-less transition. Such transitions are very common in case of solids where the excess energy is released as phonons associated with vibration of the lattice. Liquid too presents conditions conducive for relaxation of excess energy in the nonradiative manner. In the case of gas, wherein the atoms or molecules are barely in contact, there is hardly any possibility to dissipate the excess energy nonradiatively.

Fig. 7.2 Schematic
illustration of a laser based
on the four-level energy
configuration

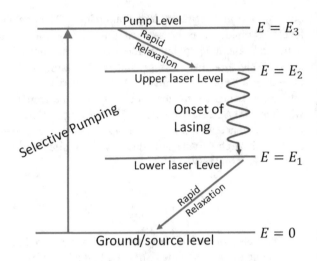

Fig. 7.2 Schematic illustration of a laser based on the four-level energy configuration

upper laser level now results in the realization of population inversion. This practically eliminates the loss of energy through fluorescence making the operation of the laser conducive for the *cw* mode.

Although the removal of loss due to fluorescence makes the process of lasing energy efficient, the maximum power that can be extracted from this laser under *cw* operation is, however, limited for the following reason: An insightful inspection of Fig. 7.2 clearly reveals that for the extraction of each laser photon of energy $E_2 - E_1$, a pump energy of value E_3 is required to be expended in the system. It is equivalent to saying that with every photon that the laser emits, an amount of energy of value $[E_3 - (E_2 - E_1)]$ stays back into the lasing medium that eventually is realized as heat. The mechanism of the generation of heat varies from laser to laser and shall be described when we study specific laser systems in Chaps. 10 and 11 of this volume. Extraction of more *cw* power from the laser will be possible only if the input pump power is also accordingly raised and consequently more heat will be liberated into the medium raising its temperature. This will result in a higher thermal population of the lower laser level and, in turn, a stunted incremental rise in the population inversion. Thus, an increased output power may be obtained initially that would certainly not be commensurate with the enhanced pump power. Eventually, a further increase in input will have a detrimental effect on the overall power that the laser is capable of giving out.

7.3 Pulsing the Laser: A Possible Remedy

We thus find that a laser based on the three-level energy scheme is not compatible for operation in the *cw* mode. Although continuous operation is possible for a four level-based laser, the maximum achievable power is, however, limited here. These

Fig. 7.3 Output power of a
cw and pulsed laser in time.
By appropriately controlling
the input energy of the pulse
and judiciously spacing the
pulses from one another in
time, the peak power of the
pulsed laser can be made
much higher

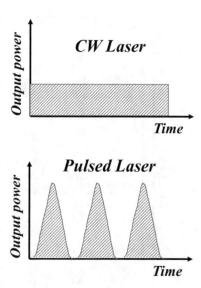

problems can be readily overcome by operating the laser in pulse mode which also
allows scaling up the laser power achievable from a four-level laser in *cw* mode.

It is possible to pulse the laser by concentrating the pump energy in time. The
power output of the laser, over a brief period of time, can thus be raised to a level
significantly higher than what is possible with *cw* operation as depicted in Fig. 7.3. In
the following section, we shall study the popular techniques of operating a laser in
pulse mode. Transfer of the stored energy into the lasing medium within a short time
results in a rapid buildup of population inversion setting up the emission of the laser
in the form of a pulse. The peak power of the pulse can be increased by increasing
both the amount of pump energy and the rate at which it is deposited into the lasing
medium. Obviously if more energy is pumped in a shorter time, population inversion
will build up more rapidly pushing the peak power further up. Scaling up the power
of the laser pulse, however, cannot occur indefinitely. The dissipation of larger
energy will result in an increased heating of the active medium that would take a
longer time to return to the ambient condition. The arrival of the next pump pulse
before this happens will understandably have a detrimental effect on the achievable
peak power. By judiciously spacing the pump pulses in time, the power of the pulse
can be progressively raised, and the illustration of Fig. 7.4 clearly reveals this
situation. It depicts the cases with two different rates at which the laser is pulsed.
In the first case (Fig. 7.4A), pulses arrive at a relatively faster rate meaning that the
laser operates here at a higher repetition rate[2] compared to the second case
(Fig. 7.4B). Furthermore, the pump energy gradually increases from the top to
bottom traces of Fig. 7.4A. In the top trace where laser power as a function of

---

[2]The number of pulses that a pulsed laser gives out in every second is called its repetition rate or
pulse repetition frequency (prf) and is expressed as Hz. Today we have lasers with prf ranging from
several Hz to the range of MHz.

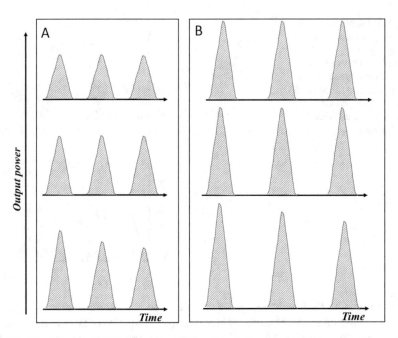

Fig. 7.4 The dependence of peak power of the pulse on the energy of the pump pulse and its repetition rate

time is shown for three consecutive pulses, the energy deposited into the lasing medium within a pulse was small enough to enable the medium to return back to the ambient condition following lasing before the arrival of the next pump pulse. Consequently, as seen, the pulses are emitted in a consistent manner in this case. In the second case, illustrated in the middle trace, the energy of the pump pulse is increased to a moderate value, and the laser continues to behave in a reliable manner at this repetition rate too. In the third case, shown in the lower trace, the pump energy is raised to a level where accumulation of heat after each pulse results in a gradual buildup of the temperature and, in turn, deterioration of the performance of the laser. The restoration of the performance of the laser is possible by appropriately increasing the interval between the two successive pulses to ensure return of the medium to the ambient condition. The laser now understandably operates at a lower repetition rate, and its performance with progressively increasing pump energy is depicted in the traces of Fig. 7.4B. Notably, the input pump energy to the laser is the same for the lower trace of Fig. 7.4A and the upper trace of Fig. 7.4B. As seen, with increasing pump energy, the accumulation of heat eventually deteriorates the performance of the laser (middle and bottom trace of Fig. 7.4A) even at this reduced repetition rate. It thus appears that the detrimental effect on the performance of the laser with increasing pump energy can be offset by a correspondingly progressive reduction of the repetition rate of operation. It is natural to expect that employing a mechanism to remove surplus heat will allow an increase in both the pump energy and repetition rate. The obvious question that arises here is: can the output of the pulsed laser, in particular its peak power, be increased indefinitely? In the following section, we attempt to answer this question by taking a deeper look at the temporal growth of the output pulse in the conventional operation of a pulsed laser.

7.4   The Maximum Achievable Power from a Conventional Pulsed Laser

In a pulsed laser, as we know now, the energy from an extraneous source is pumped into the lasing medium over a finite length of time. This causes the population inversion to gradually build up and attain a maximum value. Left to itself, the inversion will decay in time normally through spontaneous emission. In conventional operation, this active medium, as shown in the inset of Fig. 7.5, is placed inside a cavity of which one mirror (M_1) is 100% reflective and the other mirror (M_2) is provided with a transmission loss to allow coupling out of the laser beam. The rate of building up of population inversion and, in turn, the peak power of the pulse achievable from the laser will obviously depend on the rate at which input energy is pumped into the system. Here, we consider three different pumping rates, viz., small, moderate, and extremely rapid, and examine their impact on the formation of the respective output pulses of the laser. All these three situations are qualitatively

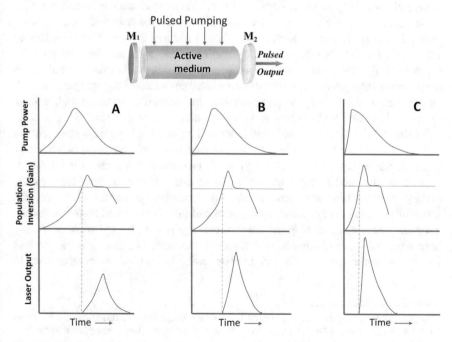

Fig. 7.5 The dependence of the building up of population inversion and laser pulse in time on the rate of deposition of pump energy is shown for three different pumping conditions: A, slow rate; B, moderate rate; and C, maximum rate that is possible to realize

represented in frames A, B, and C of Fig. 7.5. Each of these frames comprises three traces each. The upper traces depict the temporal shape of the pump pulse, the middle traces show the corresponding buildup of population inversion or gain in time, and the bottom traces reveal the formation of the laser pulse in time. The transmission loss of the cavity is obviously independent of time and therefore can be represented as a straight line parallel to the time axis as shown in the middle traces of all the three frames. At a slow rate of pumping, the population inversion (or gain) also builds up relatively slowly. Commencement of lasing, however, occurs only at the instant of gain crossing the loss line, since amplification of light, as we already know, is not possible unless gain exceeds the cavity loss. It is well known that although a stimulated emission adds one photon to the system, it also reduces the population inversion by two. The onset of stimulated emission thus spices up the situation as there now exist two competitive processes: While the act of pumping attempts to push the gain up, the stimulated emission, on the other hand, pulls it down. As the density of photons rapidly builds up inside the cavity, the rapidly increasing rate of stimulated emission would soon outweigh the effect of pumping and eventually will succeed in pulling the gain down to the loss line. Thereafter the gain will remain clamped[3] to the loss line until the gradual waning of the pump will make it finally fall below the loss line. This will essentially put a stop on any further amplification of the intracavity light. However, the photons that were already present inside the cavity shall continue to emerge through the output mirror over a time governed essentially by the cavity decay rate.[4] This completes the formation of the pulse and, in turn, results in the cessation of lasing until the arrival of the next pump pulse. The peak power achievable here will obviously depend on the maximum value of gain that can be eventually realized. The gain can be increased by augmenting the rate of deposition of pump energy as this would allow neutralization of the detrimental effect on gain by the stimulated emission for a longer period. With an increased rate of rise of the pump energy, the population inversion and, in turn, the gain will attain a higher value, and consequently the peak power of the laser pulse will exhibit a rise. The cases with moderate and rapid pumping are illustrated in the traces of frames B and C of Fig. 7.5, respectively, to corroborate this point. Clearly therefore the peak power of the laser pulse can be increased by increasing the rate of deposition of stored energy into the lasing medium. However, how fast can the energy be pumped into the medium? There is obviously a practical limitation on the speed of pumping energy into a real system and this, in turn, will therefore restrict the peak power achievable in the conventional operation of a pulsed laser. As we have seen, the onset of stimulated emission at the instant of gain crossing the loss line counters the rise of gain and thereby limits the achievable peak power. In

---

[3]The phenomenon of gain clamping, which has a strong bearing on the operation of a laser and known to manifest into a host of interesting effects, will be looked into greater detail in the next chapter.

[4]Cavity decay rate is essentially the rate at which photons escape from an optical cavity. A cavity with higher Q will decay at a lower rate as it can retain photons for a longer time and vice versa.

Chap. 9, we shall study how the onset of stimulated emission can be delayed allowing the peak power to rise much beyond what is possible otherwise in the normal operation of a laser.

7.5 Continuous and Pulsed Pumping of a Laser

Pumping, which sets off the process of lasing, is an act of transferring energy from the source to the lasing medium. As we have seen earlier in Chap. 5, Section 5.2.2, there exist a variety of techniques to effect this transfer of which some are more popular and some are more exotic. The two most popular pumping schemes, viz., optical and electrical, were discussed there basically to underline the physics of causing excitation without any reference to the *cw* or pulsed operations of the laser. We shall elaborate here on the implementation of these techniques in both *cw* and pulsed systems. Furthermore, as revealed by Table 7.1, the appropriateness of a pumping scheme critically depends on the state of the lasing medium, viz., solid, liquid, or gas. As also seen here, the pumping geometry too differs from case to case. Basically, there exist two pumping configurations, side and end. In side pumping, the pumping species, photons or electrons, traverses the lasing medium laterally, i.e., perpendicular to the optical cavity, while they travel longitudinally in the case of end pumping. Both of these arrangements are schematically elaborated in Fig. 7.6.

7.5.1 Optical Pumping

Optical pumping that basically converts light of a shorter wavelength into light of a longer wavelength was made use of by Maiman when he operated the first ever laser in 1960. The pumping photons can obviously be derived from the two types of light sources, conventional or laser.

7.5.1.1 Optical Pumping by a Conventional Source

The pump photons are normally obtained here from either an arc lamp or a flash-lamp. The basic principle of operation of such lamps involves conversion of electrical energy into light and is similar to the operation of a fluorescent light source that has been described earlier in Chap. 4. A high current discharge is usually realized in a noble gas at high pressure (~atm) contained in a quartz glass tube. In the case of an arc lamp, the discharge is sustained in a continuous manner in the form of an arc, while a glow discharge is effected for a short time in the case of a flashlamp. Clearly, therefore, an arc lamp serves as the pump source for an optically pumped *cw* laser, while a flashlamp, on the other hand, energizes an optically pumped pulsed laser.

Table 7.1 Suitability of a pumping scheme vis-à-vis the state of the lasing medium

Pumping scheme	Optical (conventional light)						Optical (laser)						Electrical					
Mode of pumping	cw			Pulsed			cw			Pulsed			cw			Pulsed		
State of the lasing medium	Sol	Liq	Gas	Sol	Liq	Gas	Sol	Liq	Gas	Sol	Liq	Gas	Sol	Liq	Gas	Sol	Liq	Gas
Suitability of the scheme	Yes	Yes	No	Yes	Yes	No	Yes	Yes	Yes	Yes	Yes	Yes	Yes	No	Yes	Yes	No	Yes
Pumping geometry	Side	Side	---	Side	Side	---	End (a)	End (a)	End	End (a)	End	End	End & side	---	End & side	End & side	---	End & side

[a]This is the popular geometry and side pumping is rare

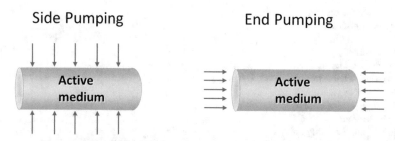

Side Pumping

End Pumping

Fig. 7.6 Schematic illustration of the two possible ways of introducing the pump light into the active medium. In the case of side pumping, they enter the medium laterally, and in case of end pumping, on the other hand, they flow in through the end faces. In the case of electrical pumping, current flows longitudinally for end pumping and transversely for side pumping

Fig. 7.7 Operation of a cw laser optically pumped by an arc lamp source. The electrical circuit for initiating and sustaining the high current arc discharge in the arc lamp is shown as well. The cylindrical enclosure is to reflect the escaping light back into the active medium

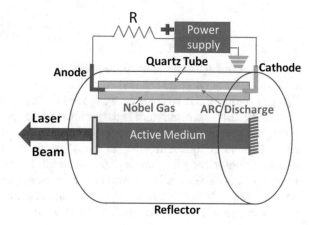

7.5.1.1.1 *cw* Lasers

The operation of a typical arc lamp pumped *cw* laser is shown schematically in Fig. 7.7. The application of an appropriate voltage between the two electrodes, fixed at the two ends of a quartz tube, strikes an arc discharge in the noble gas contained at a high pressure in the tube. Initial boosting of the applied voltage through some electrical manipulation, not shown in the figure, is a prerequisite, as discharge impedance at a high gas pressure is too large initially for just the power supply voltage to take over. The arc can be viewed as a bright filamentation along the center of the tube, depicted in this figure as a thick white line joining the anode and cathode. Owing to the nonohmic nature of the impedance of a gaseous discharge, a real resistance R is usually connected in series with it to stabilize the discharge. The filamentation of the discharge results in greatly enhancing the density of the current flowing through it. This, in turn, causes the tube to emit light spontaneously all around it of intensity much higher than is possible with a conventional fluorescent light, making the arc lamp suitable for pumping a *cw* laser. The lamp, of nearly the same length as the active medium, is obviously placed in its immediate vicinity. To

Fig. 7.8 Operation of a
pulsed laser optically
pumped by a flashlamp
source is shown
schematically. The details of
the electrical scheme to store
energy into a capacitor and
its subsequent transfer into
the interelectrode gaseous
volume of the flashlamp are
also shown

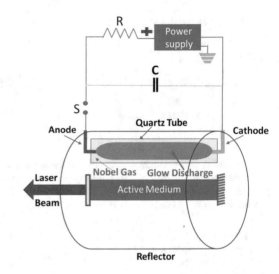

catch the escaping photons and put them back into the lasing medium, the lamp and
the laser are often enclosed in a hollow cylindrical reflective cavity. By virtue of the
high gas pressure, the lamp emits light over a wide spectrum of wavelengths of
which only a fraction contributes to the pumping of the species from the ground state
of the lasing medium. Light of longer wavelengths detrimentally impacts the operation of the laser by contributing to the origin of thermal lensing.[5] In order to
enhance the efficiency of pumping and minimize this effect, the inner surface of
the enclosure is often coated to reflect shorter wavelengths and absorb or transmit the
longer wavelengths. The use of multiple arc lamps to pump a single lasing medium is
also not uncommon as will be seen later in Chap. 10.

7.5.1.1.2 Pulsed Lasers

The operation of a typical flashlamp pumped pulsed laser is schematically shown in
Fig. 7.8. The coupling of the pump photons emerging from the flashlamp into the
lasing medium is qualitatively similar to that in the case of an arc lamped pumped *cw*
laser. The mechanism of energizing the flashlamp, however, significantly differs
from the *cw* case owing to the need for depositing electrical energy into it in the form
of a pulse. In the conventional operation, a high voltage power supply first charges
up a condenser *C* through a charging resistance *R*. Once the condenser has been

---

[5]Thermal lensing effect usually occurs in the operation of high-power solid-state lasers. The
fraction of the pump energy that stays back in the system, as we know, is often dissipated as heat
into the lasing medium. In the case of a solid-state laser, the laser rod, as a result, stays warmer at the
center compared to the periphery. This inhomogeneous cooling of the rod establishes a refractive
index gradient laterally across it that, in turn, significantly modifies the propagation of the
intracavity Gaussian beam degrading greatly the performance of the laser.

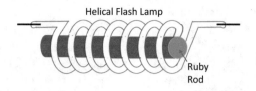

Fig. 7.9 A flashlamp with a helical geometry has an inbuilt ability of efficacious pumping when the laser rod is located at the center of the helix. The lasing medium then can be pumped from all sides, and a similar arrangement was used in the operation of the very first laser

charged to an appropriate voltage, a high voltage high current switch[6] (S) placed in the discharge loop containing the capacitor and the flashlamp is made to close. The closure of the switch, in turn, allows the capacitor to discharge its stored energy quickly into the flashlamp load initiating the flow of a pulsed current through it. This would set off the spontaneous emission of photons, enough in numbers within a short time, capable of deluging the lasing medium. The transient nature of this method of pumping brings about the realization of population inversion over a finite length of time, unlike the case of *cw* lasers where the enduring pump allows inversion to persist. All of these successive events, triggered by the closure of the flashlamp switch, eventually culminate in pulsed emission from the laser. As in the case of *cw* operation, the efficacy of pumping can be increased by the usage of more than one flashlamp for pumping a single laser. Using a flashlamp of helical geometry and locating the lasing medium at the center of the helix are also not uncommon. We now know that the very first laser made use of a spring-shaped flashlamp. By locating the ruby rod at the center of the spring, it was possible to pump it from all sides (Fig. 7.9).

As summarized in Table 7.1, optical pumping by a conventional source that emits photons in all possible directions is only suitable for solid and liquid lasers and not gas lasers. This is because here, the density of photons emitted in any particular direction is always much too low, necessitating the pumping of the lasing medium through its side, called transverse pumping, which allows interception of a substantial number of photons by it (Fig. 7.10). Effective utilization of these photons is possible only in the case of a solid or liquid medium wherein the absorbing species such as atoms or molecules are densely stacked. In the case of a gaseous medium, on the other hand, the vast empty spaces between the absorbing species allow the majority of the photons to escape rendering the task of achieving population inversion unfeasible. In the case of pumping of the laser through its ends, called

---

[6] A switch is a device that can make or break the connection of an electrical circuit. A high voltage high current switch, usually a spark gap or a thyratron, that can hold a voltage of several kV to tens of kV and capable of withstanding pulse current in the range of kiloampere is a key component in the operation of a pulsed laser.

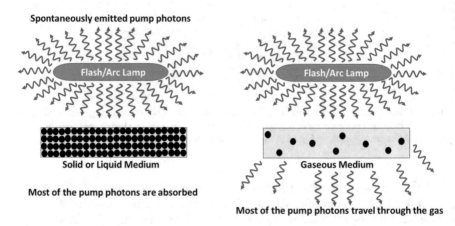

Fig. 7.10 2D longitudinal view of the lamp and lasing medium. Pump photons are spontaneously emitted in all possible directions from the lamp. Side pumping allows a large fraction of these photons to be intercepted by the lasing medium. In case of solid or liquid, most of these photons get absorbed by the species that are tightly packed. In case of a gas laser, majority of these photons escape unabsorbed

longitudinal pumping, the realization of population inversion will be a far cry even for a solid or liquid medium let alone gases! This fact has been made amply clear in the illustrations of Fig. 7.11.

7.5.1.2 Optical Pumping by a Laser

A gas laser, thus, cannot be optically pumped by a conventional source of light that emits photons all around it. A laser of appropriate wavelength, however, can work as a light source to effect optical pumping of a gas laser. The use of a laser to pump another laser also greatly enhances the energy transfer efficiency from the pump source to the active medium. The typical broadband emission of a flash or an arc lamp has a poor overlap with the width of the absorption line of the lasing medium. As illustrated in Fig. 7.12, this results in a significant fraction of the pump light, lying on either side of the absorption line being wasted. In contrast, a judiciously chosen laser with its characteristic narrow spectrum has a much closer match with the absorption feature of the active medium. Use of lasers as a pump source thus offers better utilization of the pump photons and consequently a greatly improved performance of an optically pumped laser. Furthermore, it may be noted that while a gas laser cannot be pumped by a conventional light source, the same can be achieved readily by using a second laser as the pump source.

It is worth examining here as to how the hurdles posed by a gaseous medium, as elucidated through the illustrations of Fig. 7.11, can be surmounted when a laser is used as the pump source. Unlike an arc or a flashlamp, the laser gives out a beam of light and hence fits the bill of pumping the active medium longitudinally through one

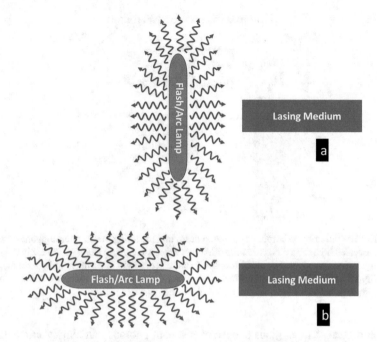

Fig. 7.11 In case of end pumping, irrespective of whether the lamp and laser are orthogonal (**a**) or in-line (**b**), only a negligible fraction of the pump photons shines onto the medium. The possibility of realization of population inversion is impractical here

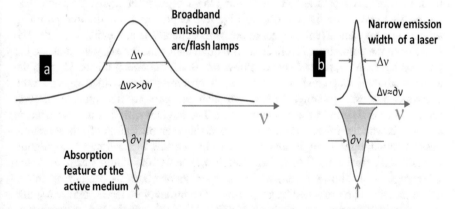

Fig. 7.12 (**a**) The width of the emission of a conventional pump source far exceeds the absorption width of the active medium. A significant fraction of the pump photons therefore stays unutilized resulting in a wasteful expenditure of energy. (**b**) A laser in contrast has a narrow spectral width and when used as a pump source can outperform the flash or arc lamps

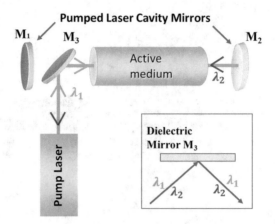

Fig. 7.13 The difficulty to introduce the pump beam into the lasing medium through one of its end faces is revealed here. The dielectric mirror M_3 that reflects the pump beam of wavelength λ_1 onto the lasing medium barely transmits the amplified spontaneous light at wavelength λ_2 precluding the possibility of lasing. The inset shows that an ordinary dielectric coated reflective surface doesn't distinguish between two not so widely different wavelengths

of its end faces. The copious supply of resonant photons can simply inundate the absorbing medium, thus providing the prospect of realization of population inversion even for a gaseous active medium. There exists, however, a practical difficulty to introduce the pump beam through one of the end faces of the active medium placed between two parallel mirrors forming the pumped laser cavity. This point can be readily appreciated by referring to Fig. 7.13 depicting a longitudinal pumping scheme that may mistakenly appear to be the most plausible in the first instance. The beam from the pump laser is reflected off a nonmetallic mirror M_3, placed inside the pumped laser cavity, formed by the output mirror M_1 and metal mirror M_2, into the active medium through one of its end faces. The pump photons will always be more energetic or possess shorter wavelength than the photons that emerge through spontaneous emission in the active medium. The obvious question that we need to ask here is whether intracavity mirror M_3 will allow amplification of this spontaneously emitted light even if this end pumping causes realization of population inversion. To make the illustration readily palatable, the pump beam of lower wavelength λ_1 is shown green, while the spontaneous light λ_2 originating in the active medium is represented by red arrows. The ordinary dielectric mirror M_3 that reflects light at the pump wavelength λ_1 will also significantly reflect light originating through spontaneous emission at a slightly longer wavelength λ_2. This fact has been exemplified in the inset of this figure. The spontaneous photons emitted longitudinally along the cavity in either direction will grow in number as they travel through the gain medium. However, mirror M_3 reflects all these photons arriving from the active medium toward the pump laser rendering M_1 nonfunctional.

Clearly the scheme will work if, as shown in Fig. 7.14, the normal dielectric mirror is replaced by a special mirror called the dichroic mirror that offers high reflectivity at one wavelength and, high transmission, at the same time, to another

Fig. 7.14 Use of a dichroic mirror to couple the pump beam into the lasing medium makes this pumping scheme feasible. The inset depicts operation of a typical dichroic mirror

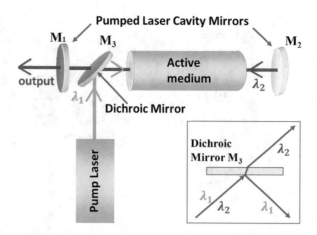

wavelength. The operation of such a mirror is elucidated in the inset of this figure. The dichroic mirror M_3 allows to and fro journey of the light at wavelength λ_2 in the pumped laser cavity leading to the onset of lasing. Thus, we see that operation of an optically pumped gas laser becomes feasible when it is pumped by another laser. The longitudinal pumping scheme, which is mandatory here, will also work for solid and liquid lasers. The relatively low optical damage threshold of the dichroic optics, however, limits the maximum achievable power of the laser in this pumping scheme. We, having studied the behavior of light (Chap. 2), now know that the reflection of light depends critically on its polarization. This property can be exploited to achieve a pumping scheme that would allow operation of such a laser by dispensing with use of dichroic optics, significantly enhancing the maximum achievable power. The realization of such a scheme, however, calls for a linearly polarized pump laser. A laser in its conventional operation emits randomly polarized light. As we shall see in the following section, the Brewster principle, already elaborated in Chap. 2, can be exploited to construct a laser capable of giving out polarized light.

7.5.1.2.1   Manufacturing Polarized Light from a Laser

The optical scheme to make a laser switch its emission from the random to plane polarization is illustrated in Fig. 7.15. The heart of this scheme is an ordinary optical element, transparent at the lasing wavelength that is placed at the Brewster angle of incidence, θ_B, between the active medium and one of the cavity mirrors. Brewster's principle has been described earlier (Sect. 2.8.1), and the inset of this figure will help recapture the same. The spontaneous photons emitted by the medium, immediately following its pumping, act as the seed for the stimulated emission to set in. As spontaneous light is essentially randomly polarized so would also be the light emitted through stimulated emission in the beginning. The random polarization, as we have seen in Chap. 2, can be resolved into two orthogonal components, P and S.

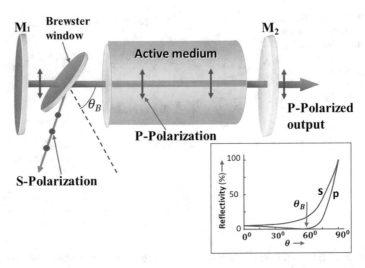

Fig. 7.15 The optical layout of the scheme to obtain polarized output from a laser is illustrated here. The inset of the figure shows the reflectivity of s and p polarization as a function of the angle of incidence

At the Brewster's angle of incidence θ_B, the P-polarization exhibits 100% transmission, while the S-component is partly transmitted and partly reflected. It may be noted here that for a glass substrate, for which θ_B assumes a value of ~57°, the S-polarization is split equally between the reflection and transmission as the light travels through it. As the light begins to grow through to and fro journeys inside the cavity, it would lose a part of the S-component of the polarization in each passage through the Brewster plate while its P-component stays intact. The rise in the number of photons belonging to the P-polarization will, in turn, result in their further rise through stimulation. This positive regenerative effect will cause the buildup of P-polarization so rapidly that almost instantly the cavity will be completely devoid of any S-polarization. The light emitted by the laser will therefore be linearly polarized. Rotation of the Brewster window about the axis of the laser will essentially rotate the incident plane and, in turn, the direction of polarization.

7.5.1.2.2 Optical Pumping by a Linearly Polarized Laser

The dielectric coating of the dichroic mirrors is extremely prone to optical damage and severely restricts the maximum power achievable from a laser that has a dichroic intracavity mirror. A pump laser capable of providing a polarized output beam can help overcome this limitation as it allows complete dispensation of a dichroic mirror in the operation of an optically pumped laser. The optical pumping scheme through one of the end faces of the active medium that makes use of a polarized pump laser is

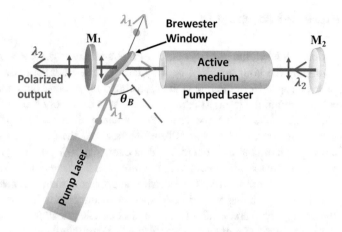

Fig. 7.16 This depicts the optical scheme to pump a laser with a polarized light beam. A fraction of the polarized output of the pump laser can be coupled into the lasing medium through reflection off a transparent uncoated substrate placed at the Brewster angle. Dispensation of the expensive and delicate dichroic coupling mirror allows a significant increase in the maximum achievable power from the pumped laser

illustrated in Fig. 7.16. A fraction[7] of the output from the pump laser, when on S-polarization, can be coupled into the active medium through reflection off a transparent optical element placed at the Brewster angle. The pumped laser can lase only on P-polarization as the Brewster window allows complete transmission of this polarization. The S-polarized light that exhibits a significant reflection loss during its passage through the Brewster window cannot compete with the P-polarized light and is completely annulled by it. The coupling of the pump beam through a Brewster plate thus offers the following advantages:

1. Dispensation of the expensive and delicate dichroic mirror.
2. Ability to generate higher output power.
3. Intrinsic ability to provide a polarized output beam.

   We have discussed a variety of optical pumping schemes that include laser and non-laser pump sources as well as side and end pumping arrangements. Needless to say, a pumping scheme that is useful for a particular laser may not perform as efficiently in the operation of another laser. The aspect of laser specific aptness of the pumping schemes will be more apparent when we study in detail the operation of a few specific laser systems in the latter chapters of this volume.

---

[7] In the case of a glass substrate of refractive index 1.5, 50% of the S-polarized light is reflected at the Brewster angle of reflection. The fraction of reflection monotonically increases with increasing refractive index of the Brewster plate.

7.5.2 Electrical Pumping

In the operation of a laser, the primary source of input energy is generally electrical. In case of pumping the laser optically, be it by laser or non-laser sources, the input electrical energy drawn from the line is first consumed to generate the photons that would eventually pump the laser. This intermediate step invariably brings down the wall-plug efficiency (WPE)[8] of the laser. For instance, if a flashlamp converts electrical energy to optical energy at 50% efficiency and then the laser converts this flashlamp optical output into coherent optical energy at 10% efficiency, then the WPE of the laser is clearly 5%. In contrast if a laser can be directly pumped by the electrical energy that is drawn from the electrical supply line, then the intermediate step of generation of pumping photons is dispensed with. This will cause a major boost in the energy conversion ability of the lasers. Electrical pumping, which basically relies on the transfer of the electron's kinetic energy into the lasing medium through collisions with its constituent atoms/molecules/ions, is not practical for liquid or solid-state lasers. One exception, as we shall see in a latter chapter, is of course semiconductor lasers. The ability of a gaseous medium, on the other hand, to present a collisional environment impromptu makes a gas laser most suitable for electrical pumping both in *cw* and pulsed modes of operation.

7.5.2.1 *cw* Lasers

The typical arrangement of electrical pumping of a *cw* gas laser wherein the energy drawn from the supply line is unloaded directly in the lasing medium is depicted in Fig. 7.17. The mechanism of transferring energy is similar to that of a fluorescent light source. The stray charges, primarily electrons, get accelerated due to the

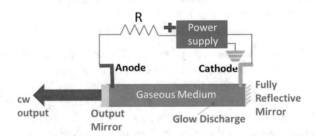

Fig. 7.17 Typical operation of an electrically pumped cw gas laser is schematically illustrated. The arrangement of the electrical circuitry for transferring energy directly into the gaseous medium from the power supply by way of creating and sustaining a glow discharge is also indicated

---

[8]Wall-plug efficiency in the present context means the efficiency with which the laser converts the total input electrical power eventually into the coherent optical output and will be dealt in greater detail in the following sections.

impressed field, gain kinetic energy, collide, and cause further ionization resulting in a glow discharge filling the entire volume of the gas enclosed in the dielectric housing. The ballast resistance[9] R is connected in series with the gaseous discharge load, which is of nonohmic nature, to stabilize the discharge. It should be noted here that such a glow discharge can be struck only when the gas pressure is low (typically within a few tens of mbar). With increasing pressure, the discharge begins to get constricted and eventually culminates into an arc detrimentally affecting the performance of the laser.

The inelastic[10] collisions that the electrons undergo with the gaseous species causing their excitation to a higher level of energy are essentially the mechanism of the electrical pumping of the lasing medium. To enhance the pumping efficiency, the operation of a gas laser often calls for mixing the lasing gas with one or more surrogate gases. This would be apparent when we take a closer look at the operation of a few specific gas laser systems in Chaps. 10 and 11 of this volume. Irrespective of the number of gaseous species, the energy drawn from the input power line first appears as the kinetic energy of the electrons before being transferred to the active medium through collisions. The laser would perform in an incessant manner as long as the drawn electrical power continues to feed the glow discharge.

7.5.2.2 Pulsed Lasers

The low operating pressure of a *cw* gas laser that is a prerequisite to realize a glow discharge limits the maximum power achievable in its continuous operation. Boosting the extractable power from the lasers obviously requires augmentation of active species realizable either by increasing the active length or by increasing the operating pressure. As increasing the length beyond a point becomes impractical, the latter approach has gained more popularity. However, at a higher operating pressure that inhibits the occurrence of a glow discharge in the *cw* mode, the laser has to be essentially operated in the pulsed mode. With increasing gas pressure, the breakdown voltage[11] of the discharge gap increases, and effecting a longitudinal discharge, as in the case of *cw* operation at low pressures, becomes impractical. This problem can be overcome by going for a transverse discharge wherein the two appropriately long and contoured electrodes are positioned face to face one above another longitudinally, maintaining a reasonable gap as shown in Fig. 7.18. This

---

[9] A ballast resistance, in the present context, is an ohmic load inserted into an electrical loop to ensure a stable operation of a gaseous discharge by limiting the value of the current flowing through the circuit.

[10] A collision between two or more objects is said to be inelastic if only momentum and not energy is conserved after the collision. In an elastic collision, on the other hand, both momentum and energy are conserved. It is of interest to note that majority of collisions that we witness in our everyday life are inelastic in nature.

[11] Breakdown voltage of an electrical gap between two conductors is defined as the minimum voltage required to be applied across it to break the gap into conduction.

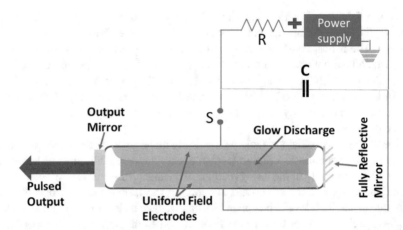

Fig. 7.18 Typical operation of an electrically pumped pulsed gas laser is schematically illustrated. The arrangement of the electrical circuitry for transferring energy directly into the gaseous medium in a pulsed mode from the power supply is also displayed

arrangement not only brings down the breakdown voltage and, in turn, the voltage requirement from the power supply but also allows the discharge to encompass significant volume of the available gas. In order to ensure a uniform field across the entire gap between the electrodes, a prerequisite for realization of a glow discharge, the electrodes are spatially contoured [23] along the perimeter to ward off the edge effect caused by the enhancement of the field at a sharp point or edge of a conductor. In the conventional operation, a high voltage power supply first charges up a condenser C through a charging resistance R. Once the condenser has been charged to an appropriate voltage, a very fast switch S, normally a spark gap or a thyratron, placed in the discharge loop containing the capacitor and the laser load, is made to close. The closure of the switch, in turn, impresses a rapidly building high voltage impulse across the gap between the electrodes leading to its breakdown. This allows the capacitor to discharge its stored energy quickly into the gaseous volume enclosed between the electrodes causing excitation of the species to higher energy levels through electronic impacts. The discharge that is short-lived brings about the realization of population inversion over a finite length of time setting off the emission of the laser in the form of a pulse. As the discharge operates here in a transverse configuration, this class of lasers is usually referred to as transverse electric gas lasers, often abbreviated as TE gas lasers. The high-pressure operation of a CO_2 laser makes use of a transverse electric discharge for its operation and will be described in Chap. 11.

We have considered here the pulsed operation of a gas laser at high pressure. At low operating pressures, where a gas discharge invariably runs in the glow mode, the laser can, of course, also be operated in the pulsed mode too. Such an operation has limited applications as it is not conducive to the generation of high power output. As we shall see later, the high-pressure operation of a TE gas laser can readily generate pulses of average power to the tune of tens of MW over a duration spanning several hundreds of nanoseconds as against typical tens of watts in the conventional *cw* operation.

7.6   Operating Efficiency of cw and Pulsed Lasers

There exist many possible ways, and rightly so, of defining the efficiency at which a
laser performs. The rationale behind these multiple approaches of expressing the
efficiency of a laser is best explained by considering the case of a typical laser, the
operation of which is based on a four-level energy scheme. To this end, we take
another look at Fig. 7.2 of Sect. 7.2 of this chapter that depicts the operation of a laser
based on such an energy level scheme. It is easy to discern that to optically or
electrically excite an atom or a molecule from the ground to the pump level, an
energy of value E_3 is required to be expended. Only a part of this excitation energy is
available to manufacture a laser photon of energy $(E_2 - E_1)$ as the remaining part is
mandatorily dissipated into the system, often as heat. Even if we consider an ideal
situation where the entire energy being drawn from the supply line is available to
produce excitations, the above fact places an upper limit on the maximum achievable
efficiency. The efficiency of such an ideal laser is termed in the normal parlance as
quantum efficiency, and clearly the operating efficiency of a laser can never exceed
this value. In reality, however, only a fraction of the energy drawn from the line is
available for production of photons, in case of optical pumping, or electrons, in case
of electrical pumping. Further, as we already know, only a fraction of these photons
or electrons participates in the process of excitation and, in turn, realization of
population inversion. All of these different processes with a direct bearing on the
ultimate operating efficiency of a laser are represented individually in the block diagram shown in Fig. 7.19 and form the basis of the multiple nomenclatures to label
the efficiency of lasing. Clearly the entire energy drawn from the supply line, viz.,
E_{in}, cannot be directed into pumping the lasing medium. Regardless of whether it is
optical or electrical pumping, a part of the drawn energy is always lost, and the same
has been shown as dissipation on the ohmic load in Figs. , 7.8, 7.17, and 7.18. We
assume that $P\%$ of E_{in}, viz.,

$$E_P = E_{in} \times P/100 \qquad (7.1)$$

is available for the production of pump photons or electrons as the case may
be. Furthermore, all these photons or electrons do not contribute to causing excitation. We assume that $Ex\%$ of the pump energy E_P, viz.,

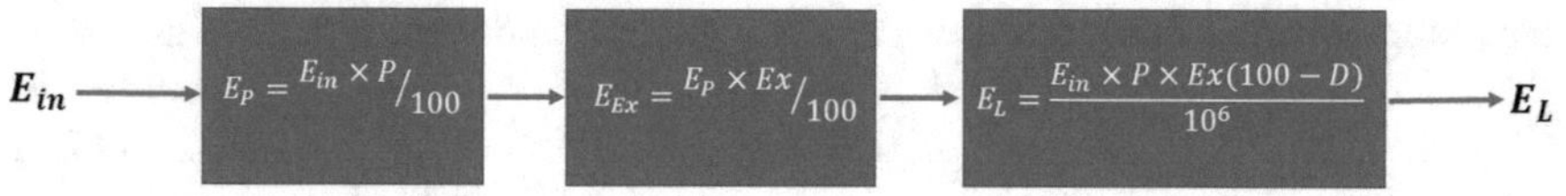

$$E_{in} \longrightarrow \boxed{E_P = \frac{E_{in} \times P}{100}} \longrightarrow \boxed{E_{Ex} = \frac{E_P \times Ex}{100}} \longrightarrow \boxed{E_L = \frac{E_{in} \times P \times Ex(100 - D)}{10^6}} \longrightarrow E_L$$

Fig. 7.19 A block diagram representation of the intermediate processes preceding lasing. A
fraction of the input energy is available for pumping (first block) of which again a fraction actually
contributes to excitation (middle block). A part of the excitation energy is mandatorily dissipated as
heat and the rest appears as the laser output (last block)

$$E_{EX} = E_P \times Ex/100 \qquad (7.2)$$

is available for causing excitation and, in turn, realization of population inversion. Understandably a part of this excitation energy is used up in the creation of the laser photons, and the rest is mandatorily dissipated into the system, often as heat. Assuming that $D\%$ of the excitation energy E_{Ex}, viz., $E_{Ex} \times D/100$, is lost due to dissipation, the energy that is eventually channeled into lasing is thus.

$$E_L = E_{Ex}(1 - D/100) \qquad (7.3)$$

Combining these three equations, we obtain.

$$E_L = \frac{E_{in} \times P \times Ex}{100 \times 100}(1 - D/100) \qquad (7.4)$$

The efficiency of the laser, which basically is the ratio of the laser energy to the input energy, can therefore be expressed as.

$$Efficiency = \frac{E_L}{E_{in}} = \frac{P \times Ex}{10^6}(100 - D) \qquad (7.5)$$

With this background knowledge, we are now in a position to elaborate on the multiple ways of expressing the operating efficiency of a laser.

Quantum Efficiency As we have seen, the quantum efficiency (QE) represents the operating efficiency of an ideal laser wherein the entire energy drawn is available for the creation of population inversion. This means that both P and Ex will assume a value of 100%, and consequently the quantum efficiency (QE) can be expressed from Eq. 7.5 as

$$QE = (1 - D/100) \qquad (7.6)$$

This is understandably governed only by the value of D, i.e., the loss of excitation energy through dissipation. For example, if D assumes a value of 20%, meaning 20% of the excitation energy is realized as heat in the lasing medium, then QE amounts to 80%. It is improbable for a practical laser to perform at the level of QE let alone surpass it.

Pumping Efficiency Pumping efficiency (PE) essentially is the efficiency at which the pump energy is consumed toward achieving excitation and is defined as

$$PE = Ex/100 \qquad (7.7)$$

where Ex is the percentile utilization of the pump energy in effecting excitation of the atoms from the ground to the higher energy state. In the case of optical pumping, pump energy, as we know, is first fed into the flash/arc lamp that, in turn, gives out the pump photons. Only a fraction of these spontaneously emitted photons eventually succeeds in causing excitations. Clearly, the efficiency at which the pump photons are created in conjunction with their utilization in achieving excitation will therefore determine the pumping efficiency. In the case of optical pumping with a laser, on the other hand, the pumping efficiency will be determined by combining the efficiency at which the laser performs and the utilization of the photons it gives out. In the case of electrical pumping, the pump energy, as we already know, first initiates a gas discharge that comprises an avalanche of electrons and ions and other neutral species. The pump energy is primarily transferred to the electrons as their kinetic energy, only a fraction of which is eventually realized as the energy of excitation. The pumping efficiency here will be determined by the combined efficiency of these intermediate processes eventually leading to useful excitations.

Electro-Optic Efficiency Electro-optic efficiency (EOE) is defined as the efficiency of the conversion of pump energy, which is essentially electrical in origin, into the energy of laser light. As the starting point here is the pump energy, the EOE is thus basically the product of PE and QE and can be expressed as.

$$EOE = Ex/100 \times (1 - D/100) \qquad (7.8)$$

For example, in the operation of a laser possessing a QE of 80%, if only half of the pump energy is realized as the energy of excitation, then the EOE at which the laser performs is clearly $80\% \times 50\%$, i. e., 40%.

Wall-Plug Efficiency The wall-plug efficiency (WPE) of a laser is defined as the efficiency at which it converts the energy drawn from the supply source into its light output. As the starting point here is basically the energy drawn from the line, it qualifies as the most practical way of describing the performance of a laser, viz., E_L/E_{in}

, and has already been defined in Eq. 7.5. Thus, in the operation of the above laser with 40% EOE, if the input energy E_{in} could be converted into the pump energy E_P at 50% efficiency, then the WPE of this laser will essentially be $40\% \times 50\%$,i.e., 20%.

7.6.1 cw Lasers

A *cw* laser operates continuously over a certain period of time by drawing energy from the supply line for that duration yielding coherent light over the same duration. Instead of denoting the total energy drawn or output energy yielded by the laser during its period of operation, it is more appropriate to refer to these as drawn power or the output power. Clearly, in the *cw* operation, a fraction of the power drawn from the supply line is dissipated on the current limiting resistor, and the rest is consumed in the production of photons, for optical pumping, or electrons, for electrical pumping. We also know that only a fraction of these photons or electrons are able to participate in the process leading to excitation and a fraction of this excitation energy, depending on the QE, is converted into coherent output power. All the above classifications therefore fit appropriately in the operation of a *cw* laser except for the wall-plug efficiency for the following reason. The fraction of the excitation energy that stays back in the system rapidly builds up as the operation of the laser continues nonstop. This would result in heating up of the lasing medium, detrimentally impacting its performance unless a heat exchanging mechanism for removal of the accumulated heat is incorporated. The operation of the heat exchanger, which calls for drawing additional power from the supply line, will understandably reduce the wall-plug efficiency of the laser. The pumping and electro-optic efficiency of the laser, however, shall remain invariant.

7.6.2   Pulsed Lasers

It is well known that a pulsed laser can operate in two different modes, viz., repetitive and single shot modes. As the name suggests, the laser is repeatedly pulsed in the repetitive mode, and the performance of the laser for a given pulse is impacted by the residual heat originating from the preceding pulse. Incorporation of a heat exchanger, as in the case of *cw* operation, will have an enduring effect on the pulse-to-pulse reproducibility in the repetitive operation. In the single shot mode of operation, on the other hand, the two successive pulses are spaced enough in time to ensure that the residual heat originating from the preceding pulse is conducted away before the arrival of the next pulse. These aspects have been elaborated vis-à-vis the operating efficiency of the laser in the following sections.

7.6.2.1   Repetitive Operation

The temporal output of the successive pulses in the repetitive operation of a pulsed laser is schematically depicted in Fig. 7.20 with (lower trace) and without (upper trace) a heat exchanger. The pump energy and all other parameters have been chosen to yield the most optimized performance in terms of energy conversion efficiency. There are three key output parameters, viz., the peak power, the intra-pulse average power, and the inter-pulse average power, that basically describe the performance of a pulsed laser. In the absence of any mechanism to remove the accumulated heat from the system, the efficiency of the laser, both electro-optic and wall-plug, exhibits monotonic reduction in time, a fact revealed by the gradual lessening of all three pulse parameters. The pulse-to-pulse reproducibility will be ensured when a heat exchanger of appropriate capacity is employed, as in the case of a *cw* laser, to prohibit accumulation of heat. This is manifested in the invariance of the pulse parameters as the operation of the laser continues. Although the additional power drawn from the supply line and consumed by the heat exchanger will have a negative

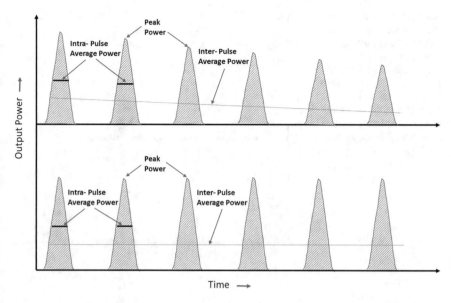

Fig. 7.20 The output power of the successive pulses in the repetitive operation, without heat exchanger (upper trace) and with heat exchanger (lower trace). The intra-pulse and inter-pulse average power are also shown in both cases

effect on the wall-plug efficiency of the laser, the electro-optic efficiency, by definition, will not be affected by this. With increasing repetition rate, the laser would perform reliably, if the heat exchanging capability is also appropriately augmented, albeit at a further expense of the wall-plug efficiency.

7.6.2.2   Single Shot Operation

As we know, in the single shot mode of operation, the performance of the pulsed laser is not affected by the residual heat originating from the preceding pulse. The act of pumping, which triggers the process of lasing, commences only after allowing enough time to ensure that the medium has returned to the ambient condition following the termination of the preceding pulse. The operation of a single shot laser is represented in the traces of Fig. 7.21 with a progressively increasing value of the pump energy. With increasing pump energy, the interval between two successive pulses is correspondingly increased to preserve the criteria of single shot mode operation. Unlike in the case of repetitive operation, the residual heat from the preceding pulse is dissipated away through natural processes. This means that for a given value of pump energy, the wall-plug efficiency in the single shot operation will always exceed that in the repetitive operation although the electro-optic efficiency will remain unchanged.

Fig. 7.21 Temporal output of two successive pulses of a pulsed laser in the single shot operation is shown. With increasing energy of pumping as more heat stays back in the lasing medium, the arrival of the next pulse is also increasingly delayed to ensure return of ambient conditions

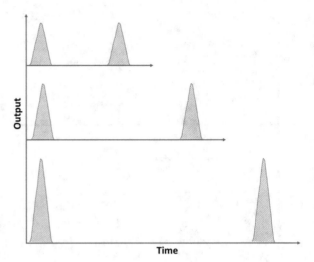

Chapter 8
Broadening of Gain and Its Bearing on the Laser Subtleties

8.1 Introduction

Spontaneous emission, as we now know, provides the seed photons for the onset of stimulated emission and understandably therefore is expected to have a major say on the emission feature of the laser itself. It is also well known that a fluorescent optical source derives its light from spontaneous emissions and that such a source gives out multicolored light. It is worthwhile at this point to examine the origin of this polychromatic emission. To this end, we revisit the operation of a fluorescent light source elaborated in Chap. 4. Upon undergoing inelastic collisions with the electrons abundantly present in the discharge, the ground state atoms climb to the excited state. Nature has always a preference to preserve a minimum energy configuration, and consequently the excited atoms spontaneously return back to the ground state releasing the excitation energy as a quantum of photons. These photons emitted randomly in all possible directions constitute the output of the fluorescent light. The situation is schematically represented in Fig. 8.1. If the excited level, located at an energy of E_1, were monoenergetic, then the spontaneous light emitted from the transition between this and the ground level would have a fixed frequency E_1/h, called the line center frequency ν_0. Consequently, the resulting fluorescence would have been monochromatic with a corresponding wavelength of c/ν_0! Where lies the catch then? An excited energy level is of course unstable, but that does not mean that an electronically excited atom has to decay at once to the ground state. The excited state, as a matter of fact, is short-lived, and that holds the key to the polychromatic nature of fluorescence. A finite lifetime τ associated with an excited energy state, according to Heisenberg's uncertainty principle, results in its spreading symmetrically on either side of the line center, and the extent of spread is inversely proportional to τ. The ground state being stable has an infinite lifetime and hence would not experience any spread as is also seen in this figure. Clearly, transitions originating from increasingly above E_1 will fluoresce light of frequency ν_0^+ progressively higher

D. J. Biswas, *A Beginner's Guide to Lasers and Their Applications, Part 1*,
Undergraduate Lecture Notes in Physics,
https://doi.org/10.1007/978-3-031-24330-1_8

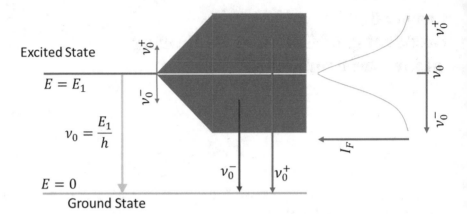

Fig. 8.1 The excited energy state has a finite lifetime, and as a consequence of Heisenberg uncertainty principle, the energy level spreads into a band. This gives rise to a corresponding spread in the frequency or wavelength of the fluorescence light. The intensity of fluorescence (I_F) exhibits a peak at the line center frequency ν_0 and gradually drops off on either side

than ν_0. Likewise, transitions arising from gradually below the central energy will yield light of frequency ν_0^- progressively lower than ν_0. The intensity of fluorescence I_F, if monitored as a function of frequency, will reveal a spectral broadening about the line center frequency ν_0, an irrefutable signature of the polychromatic nature of the fluorescence.

The aforementioned lifetime that is governed by the radiative decay of the excited energy state is denoted as radiative lifetime. There are other parameters, such as the gas pressure and temperature that will also have a strong bearing on the lifetime of an excited state and will be discussed later in this chapter. However, the point that we need to underline here is as to how this broadening of energy level, to which fluorescence owes its polychromatic nature, will impact the emission feature of a laser? As we know, inverting the population between two energy levels is central to the amplification of light by stimulated emission of radiation. Now that the excited energy state is broadened, the population inversion and, in turn, the gain does not occur on a single frequency and instead is distributed over the entire frequency spread. This fact has been qualitatively illustrated in Fig. 8.2, which, for the sake of easy palatability, depicts two situations, namely, the hypothetical case of infinitely long-lived excited state and the real case of an excited state with finite lifetime. In the former case, as shown in the top trace of this figure, the entire population inversion and, in turn, the gain would build up right at the line center frequency, viz., ν_0. In reality, however, as shown in the lower trace, the excited state is short-lived and consequently the energy level is broadened. The population inversion or gain is now distributed over the entire frequency spread across which the excited medium is known to fluoresce. The peak value of the gain, viz., g_0, understandably occurs at ν_0 as the intensity of fluorescence also peaks here. The distribution function depends on the nature of gain broadening and will be described later. Based upon the mechanism that controls the lifetime of the excited state, the broadening of gain can be categorized as either homogeneous or inhomogeneous. Needless to say, both

Fig. 8.2 The distribution of gain as a function of frequency of emission. Top trace – hypothetical case of excited state with infinite lifetime – gain exists only at the line center. Bottom trace – in reality, an excited state is short-lived, and gain is broadened in frequency

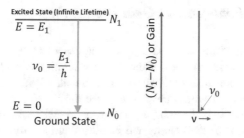

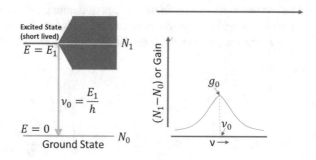

the broadening of gain and its nature affect the dynamics of a laser in a distinct manner. In addition to providing deeper physical insight into the two kinds of gain broadening, this chapter also reflects the profound impact of it on the operation of lasers. Recounting the gainful exploitation of the indelible trails it leaves on the laser dynamics will be yet another major endeavor of this chapter.

8.2 Nature of Gain Broadening

Broadening of an optical transition in frequency owes its origin to the fact that an excited state cannot exist indefinitely. Put in another way, the excited energy level is not infinitesimally narrow, a prerequisite for emission of light of a unique frequency or wavelength. In the example of the preceding section, the radiative or spontaneous relaxation of the excited atoms to the ground level governed the lifetime of the excited state, aptly termed as the radiative, spontaneous, or natural lifetime. The lifetime here depends on the spectroscopic properties of the energy levels involved in the optical transition and can vary widely from one excited level to another. The active medium of a laser, as we know, can be of all three kinds, viz., solid, liquid, and gas. Depending on the state of the medium, there can be a variety of other effects that can also directly and decisively influence the excited level lifetime and, in turn, the broadening of gain of the corresponding lasing transition. Based on the nature of the gain broadening, it is classified as either homogeneous or inhomogeneous gain broadening.

8.2.1 Homogeneous Gain Broadening

Broadening of the gain is said to be homogeneous when every atom (or molecule or ion, as the case may be) of the active medium has exactly identical fluorescence linewidth. This means that upon excitation, all of them can emit spontaneously at a given frequency within the fluorescence profile with equal probability. This point can be readily understood by referring to Fig. 8.3. We consider an ensemble comprising N excited atoms and identify them as atom-1, atom-2, atom-3, and so on. The emission bandwidths of all these atoms are individually depicted in this figure. The probability of emission at a given frequency is the same for all the atoms. This has been shown to hold true for seven randomly chosen frequencies, viz., ν_0, ν_1, ν_2, ν_3, ν_4, ν_5, and ν_6. Clearly therefore the emission bandwidth of the laser when all these atoms are put together will be identical to any of the individual atoms. This

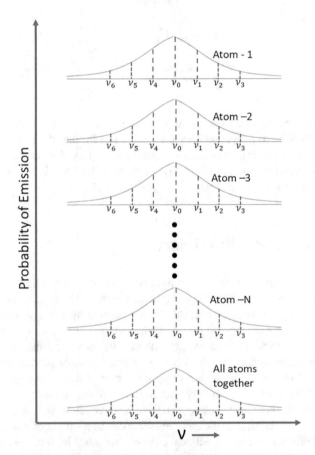

Fig. 8.3 In the case of a laser with homogeneous gain broadening, all the excited atoms emit with equal probability at a given frequency. The combined probability of emission of all the atoms is normalized to fit the scale

means that if a photon of a particular frequency, say ν_l, can interact with any one of the atoms of the ensemble, it is capable of interacting with all of them with identical probability. This fact, which holds true for photons of any frequency lying within the broadened profile, rivetingly manifests itself in the operation of a homogeneously broadened laser influencing its performance in a remarkable manner.

It may be noted here that broadening arising from atomic or molecular collisions is intrinsically homogeneous in nature. To appreciate this fact, it is imperative that we understand how the collisions result in broadening of a spectral line in the first place. To this end, we consider a gaseous atomic system that essentially presents a collisional environment. Here, the time that an atom can spend in an excited state is not governed solely by radiative decay as it can also lose the energy of excitation through collisions. Collision will thus reduce the lifetime of the excited state and consequently enhance the spectral broadening. As the probability of undergoing a collision is the same for all the atoms making up the gain medium, the broadening arising out of collision is therefore essentially an example of homogeneous broadening. Collisional broadening is also called pressure broadening as the rate of collisions increases with increasing gas pressure. As we shall see later, the pressure broadening of the gain plays a dominant role in the operation of a gas laser.

In the case of a solid-state laser, the atoms that constitute the active medium cannot, unlike the gas lasers, move around as each of them is basically tied to the host crystal's[1] corresponding lattice points. Although there is no scope of interatomic collisions, the omnipresent thermal energy invariably sets off vibrational motion of the lattice. The vibration modulates the energy levels of the atoms culminating in the broadening of its emission frequency. Such broadening is termed as thermal broadening as it originates from the thermally induced lattice vibrations. As the lasing atoms are embedded in an ordered manner in the host crystal, each of these atoms is therefore subjected to the same vibration. Consequently, thermal broadening too, like pressure broadening, is homogeneous.

The broadening that arises from the natural lifetime being governed by radiative or spontaneous decay is termed natural broadening. There is no denying of the fact that the radiative decay of the excited atoms (molecules or ions) occurs in an isotropic manner meaning they are indistinguishable as far as spontaneous emission is concerned. Thus, natural broadening, like pressure and thermal broadening, is also homogeneous in nature. It may be noted here that a homogeneously broadened gain profile, regardless of its origin, is always represented by a Lorentzian distribution function [24].

---

[1] In a majority of the solid-state lasers, the light emitting atoms are embedded in small proportion into a second species termed as the host material that is transparent to the pump and lasing wavelengths.

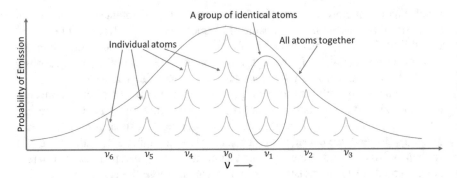

Fig. 8.4 In the case of a laser with inhomogeneously broadened gain, individual atoms emit at different frequencies. The probability of emission of the gain medium as a function of frequency is a convolution of the probability of emission of all the individual atoms putting together

8.2.2 Inhomogeneous Gain Broadening

Broadening of the gain is said to be inhomogeneous when, in total contrast to homogeneous broadening, the frequency of emission differs from atom to atom. For ease of understanding, let us first consider the operation of a gas laser. As far as emission of light is concerned, the gas atoms can be categorized into multiple groups, and each group has its own unique frequency of emission. The number of atoms belonging to any particular group also varies widely. The situation is schematically illustrated in Fig. 8.4, which captures a representative behavior of emission of individual atoms belonging to a few groups as well as the emission profile of all of them putting together. As seen, the number of atoms capable of emitting at ν_0 is maximum, and consequently the intensity of fluorescence at this frequency will also be maximum. The broadening of the fluorescence of individual groups, governed here by the radiative lifetime of the excited atoms, is homogeneous and hence Lorentzian. Clearly, the intensity of fluorescence at any given frequency is proportional to the number of atoms in the group capable of fluorescing at that frequency. The resulting inhomogeneous gain broadening of the laser is the convolution of the homogenous fluorescence broadening of all the individual groups of excited atoms and is also shown in this figure. It is of interest to note here that an inhomogeneously broadened gain profile of a gaseous medium is represented by a Gaussian distribution function [25].

The obvious question that arises here is what causes the frequency of fluorescence of one excited atom to differ from another. The answer to this can be found in the phenomenon of the optical Doppler effect. Named after the Austrian physicist Christian Doppler (1803–1853) who discovered it in 1842, our familiarity with the acoustic Doppler effect dates back to almost the beginning of the nineteenth century when steam powered locomotives began commercial operation. It is common knowledge that to a passenger waiting at a railway station, the whistle of an approaching engine appears high pitched and that of a receding engine seems low

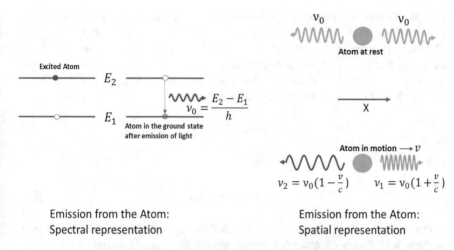

Fig. 8.5 Left: An atom excited to a state of energy E_2 comes down to the ground state of energy E_1 releasing its energy of excitation E_2-E_1 as a photon of frequency ν_0. Right: Emission from the static and moving atoms is shown spatially. The photon can be emitted randomly in any direction. When at rest, the frequency of the emitted photon is always ν_0; emissions along +ive and -ive X direction are shown here. When in motion, the frequency of the photon is blueshifted when emitted in the direction of travel and redshifted in the opposite direction

pitched. This means that when a sound producing source moves toward an observer, the observer perceives the frequency of sound to be higher. If the source, on the other hand, moves away from the observer, its frequency would appear to be lower. The faster the source moves, the greater this frequency shift.

Similar to sound waves, light waves also undergo a similar effect. To gain insight into the optical Doppler effect, we replace the sound emitting engine with a light emitting atom and examine how the frequency of light given out by a mobile atom will differ from that emitted by it when at rest. The situation is represented in the traces of Fig. 8.5. The emission of photons based on the energy level diagram is indicated in the left of this figure, while its right is a spatial representation of the emissions in the two cases, viz., when the atom is at rest and in motion. As we know, the photon can be emitted by the atom in any direction around it. However, when the atom is at rest, the frequency of the photon ν_0 is independent of its direction of travel and essentially carries the excitation energy of the atom. Thus,

$$\nu_0 = \frac{E_2 - E_1}{h} \tag{8.1}$$

where E_2 and E_1 are the energies of the excited and ground states, respectively, and h is Planck's constant. When the atom is in motion, the phenomenon of the Doppler effect comes into being, and consequently the frequency of the emitted photon now depends on the direction of emission. For the sake of simplicity, the emissions have been indicated only in two directions, along and opposite to the direction of travel of the atom. This does not mean that the emissions in the remaining directions are not important, of course they are, and as a matter of fact, we shall study the optical Doppler effect more exhaustively later in this chapter. The photon's frequency increases along the direction of atomic motion and reduces in the opposite direction and is consequently termed as blue- and redshifts of light,[2] respectively. The blue ν_1 and red ν_2 shifted frequencies can be expressed as

$$\nu_1 = \nu_0 \left(1 + \frac{v}{c}\right) \tag{8.2}$$

and

$$\nu_2 = \nu_0 \left(1 - \frac{v}{c}\right) \tag{8.3}$$

where ν_0 is the frequency of the photon that the atom emits when at rest, v is its velocity with which the atom is moving, and c is the velocity of light.

Upon acquiring this rudimentary knowledge on the optical Doppler effect, we are now in a position to take in hand the question, posed earlier, pertaining to the origin of inhomogeneous gain broadening in a gas laser. It is well known from the kinetic theory of gases that gas atoms/molecules move randomly with wide ranging velocities. The gas species are distributed within this velocity spread satisfying Maxwell's velocity distribution [26], which is critically dependent on the gas temperature. Thus, individual atoms, in the case of an atomic gas laser, move about at random velocities and emit light of a frequency, Doppler shifted by an amount dependent on the velocity. Taking another look at Fig. 8.4, some of these Doppler shifted frequencies can be readily associated with emissions at ν_6, ν_5, ν_4, ν_0, ν_1, ν_2, ν_3. As the velocity spread in the Maxwellian distribution is continuous, the Doppler shift when accounted for each individual excited atom would, in entirety, result in an inhomogeneous broadening of the laser gain. With increasing temperature, as the gas gets hotter, the faster the atoms move on average and the wider the laser gain broadening is. Doppler broadening, which is intrinsically inhomogeneous in nature, therefore increases with temperature.

It is pretty clear now that atoms belonging to any particular group of Fig. 8.4, capable of fluorescing identically at a point of time, do so as they possess the same velocity at that instant. It does therefore make sense to redraw Fig. 8.4 by portraying the fluorescence of each individual group rather than that of its constituent atoms. Figure 8.6 depicts qualitatively this new representation. As seen, the height of the fluorescence peaks that represent the intensity of fluorescence differs from group to group in line with the changing number of their constituent atoms. The linewidth of emission, however, is invariant for the following reason. Although the atoms belonging to different velocity groups fluoresce at different frequencies owing to the Doppler shift, their emission width that is governed by the excited level lifetime

---

[2]Red- and blueshift of light is the phenomenon wherein light undergoes a shift in its wavelength. The shift is red when the wavelength increases and it is blue when the wavelength decreases.

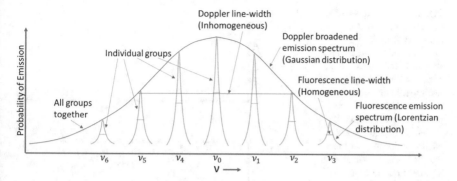

Fig. 8.6 Figure 8.4 redrawn depicting the fluorescence of each individual groups. Although the height of the fluorescence peaks differs from group to group, their linewidths remain unchanged

of the atom will remain unchanged. The convolution of the line broadening of all the individual groups represents the overall broadening of the lasing transition, which indeed is inhomogeneous and of Doppler origin. The lifetime induced broadening, as we know now, is homogeneous in nature. As long as the gas pressure is low and we conveniently ignore collisional broadening, the lifetime and the corresponding natural linewidth of an individual excited atom will be decided basically by its radiative lifetime, which is smaller than the Doppler broadening unless the gas is cooled to reduce it. At low operating pressures, the gain of a gas laser is thus predominantly inhomogeneously broadened. With increasing gas pressure, collisions begin to reduce the lifetime, and at high enough pressure, the homogeneous component of the overall broadening may outweigh the inhomogeneous part. It is of interest to note here that inhomogeneous gain broadening may also arise in the operation of certain solid-state lasers due to the presence of imperfections in the host material such as the case of an amorphous host like glass.

8.3   Gain Clamping

Gain clamping is a phenomenon that occurs when the laser gain becomes equal to the optical cavity loss and gets locked to it. Any attempt to make the gain overtake the loss by increasing the pump power would prove futile as the gain stays firmly tied with the loss line at the point of cavity mode frequency where lasing has set in. Although this may seem to contradict what has been stated earlier in many of the chapters with regard to the onset of lasing, the ensuing discussion will nevertheless help resolve the issue. For this, let us consider the *cw* operation of a homogeneously broadened laser. One obvious way to increase the gain of the laser is to increase the strength of pumping such as the magnitude of the current in the case of electrical pumping or the density of pump photons in the case of optical pumping. We, for the sake of convenience, shall confine ourselves to the case of electrical

Fig. 8.7 A typical variation of the pump current or photon density in the operation of a cw laser is shown (bottom trace). Gain as a function of frequency at few specific times beginning from the initiation of the process of pumping is shown in the middle trace where the location of the loss line is also indicated. The laser output as a function of time is displayed in the top trace

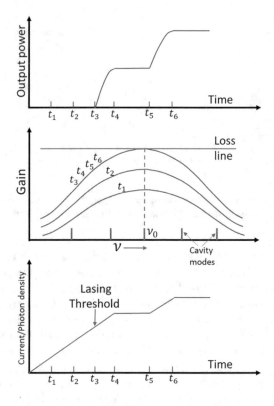

pumping alone hereafter in this chapter. Raising the current over time (Fig. 8.7, bottom trace) will also steadily increase the population inversion and, in turn, the gain. Gain, as we know, does not occur on a single frequency but instead spreads across the entire broadened, considered here to be homogeneous, lasing transition. The variation in gain with frequency is also captured in this figure (middle trace) at a few specific times beginning from the initiation of the pump. The peak value of the gain, as we know, always occurs at the line center frequency ν_0 of the lasing transition. The cavity modes that happen to lie within the gain profile have also been indicated here. For the sake of convenience, frequency of one of the cavity modes is shown to be ν_0. Cavity loss is usually independent of the frequency and thus has been represented here as a line parallel to the frequency axis. In the beginning (time t_1), the pump current is low, and consequently the gain lies much below the loss line and the process of lasing cannot obviously begin. At time t_2, the gain has risen to some extent in accordance with the increased value of the current but still lies below the loss line, and so there is no lasing yet. After a time t_3, the gain becomes high enough to touch the loss line at the line center frequency ν_0 causing the onset of lasing at this cavity mode frequency. This point has been indicated in the current vs. time curve (bottom trace) as the "lasing threshold." The plot thickens here

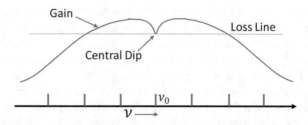

Fig. 8.8 If the cavity mode with the maximum gain, on which lasing invariably sets in, were incapable of burning the gain linked to the other frequencies, a spectral hole on the gain with a dip at the center will be formed inevitably

as the rising current attempts to push the gain beyond the loss, and the stimulated emission (SE), on the other hand, tries to pull it down.[3] With the growing density of stimulating photons inside the cavity, the rate of SE builds up so rapidly that in just a flash, the rate at which the gain builds up due to pumping equals the rate at which it is burnt out by the SE. It is equivalent to saying that the gain stays clamped to the loss line from the moment the amplification of light sets in the process of lasing. Raising the pumping strength beyond this threshold point will allow the gain to rise momentarily above the loss, and the increased rate of SE, as a consequence of the increased gain, will bring it down to the loss line practically at that moment. This means that a new equilibrium is established at a correspondingly increased power of intracavity light commensurate to the increased level of pumping. As a fraction of the intracavity light emerges as the laser beam, during the time interval t_3 to t_4 when the pump current continues to rise (lower trace of Fig. 8.7), the laser output will also exhibit a monotonic rise (upper trace of Fig. 8.7). In the operation of a laser, upon reaching the required output power, the current remains unchanged (during the interval t_4 to t_5) unless there is a requirement to boost the output. This calls for raising accordingly the magnitude of the current, and the laser output will reflect the changing pattern of current as long as the heat dissipated into the lasing medium as a consequence of lasing (and elaborated in the preceding chapter) is rapidly removed. The gain, needless to say, will remain clamped to the loss all the while.

Clearly thus, the onset of lasing at the cavity mode frequency v_0, where the homogeneously broadened gain touches the loss line first, triggers the phenomenon of gain clamping. One may begin to wonder at this point as to what prevents gain from rising on the neighboring frequencies where it is still below the loss, thereby establishing a dip at v_0 as illustrated in Fig. 8.8. The answer to this basically lies in the conduct of a homogeneously broadened gain medium vis-à-vis the emission of light. We know that every excited atom of a homogeneously broadened gain medium can emit at any frequency that lies within the domain of broadening. Returning to the situation described by the middle trace of Fig. 8.7, the gain, which exhibits a

---

[3] That each stimulated emission adds one photon to the intracavity light and reduces the population inversion by two has been touched upon in quite a few preceding chapters. Clearly, therefore, the stimulated emission causes amplification of light at the expense of the gain.

maximal behavior at ν_0, will surely touch the loss line first at this frequency setting off the process of lasing here. Being a homogeneously broadened system, gain belonging to every other frequency will also contribute to the growth of light at ν_0 ruling out any possibility of the gain touching the loss line at any frequency other than ν_0 let alone crossing it. The appearance of a central dip on a homogeneously broadened gain profile, therefore, cannot happen. (As we shall determine in a latter section of this chapter, appearance of such a dip on the gain profile, often referred to as spectral hole, is, however, inevitable in the operation of an inhomogeneously broadened laser.) It therefore appears that in the case of a homogeneously broadened laser, the mode at which lasing begins will continue to lase by burning the entire gain of the active medium ruling out the possibility of lasing on another mode. This is equivalent to saying that a homogeneously broadened laser will always lase on a single mode, a statement that, as we shall find out in the following section, is not always true.

8.4   Homogeneously Broadened Laser and Spatial Hole Burning

We have thus far confined ourselves to the spectral broadening of gain and have arrived at a conclusion that the phenomenon of gain clamping should force a homogeneously broadened laser to lase only on one mode. We intend now to delve into the possibility of a spatial effect on gain, invariably present in a linear cavity such as Fabry-Perot (FP), profoundly impacting the emission feature of such a laser, and the same is underlined in the traces of Fig. 8.9. Let us begin by considering a homogeneously broadened gain medium placed inside an FP cavity of length l. For the sake of convenience, both mirrors forming the cavity have been assumed to be fully reflective although in reality one of them will be partly leaky allowing extraction of the laser beam. The spectral profile of gain, the loss line, and a few relevant cavity modes have been indicated in trace a. As seen, the cavity mode of wavelength λ_l experiences the highest gain, and consequently lasing sets in at this wavelength. It is imperative at this point to examine as to how the onset of lasing will spatially modify the gain that initially is invariant across the entire length of the active medium (trace b). The forward and backward waves will interfere in an FP cavity to produce a standing wave[4] characterized by the presence of nodes and antinodes (trace d). The nodal planes are devoid of any radiation flux, while the antinodal planes have the maximum share of it. Clearly, the population in the antinodal regions will be eaten up by this mode, while the same will remain practically unutilized across the nodal planes. This is equivalent to saying that spatial holes will be burnt on the population inversion and, in turn, gain. These spatial holes,

---

[4]When two counterpropagating waves are superimposed, they interfere to form a standing wave interference pattern, the period of which is half the wavelength of the waves.

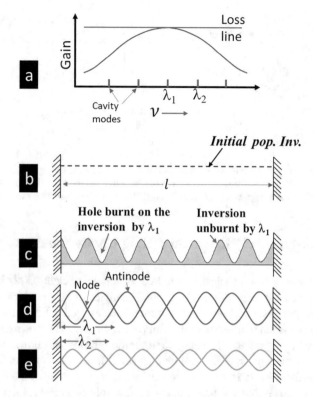

Fig. 8.9 The cavity mode of wavelength λ_1 has the highest gain (trace a). Upon placing the active medium inside a Fabry-Perot cavity of length l (trace b), lasing sets in on this mode that forms a stationary wave (trace d). The population inversion is burnt at the antinodal planes and remains unutilized across the nodal planes (trace c). An adjacent mode of wavelength λ_2 with lower gain can manage to survive by taking advantage of this unburnt gain (trace e)

occurring periodically on the population inversion (trace c), enrich the dynamics of a homogeneously broadened laser, as we shall below, in a remarkable manner. Let us consider the case of an adjacent mode of wavelength λ_2 that by virtue of possessing lower spectral gain (trace a) would, under normal circumstances, not be able to lase. However, as the standing wave pattern of one wave will have a spatial shift with respect to the other (trace d and e), the weaker mode can break into lasing by using up the gain that remained unutilized by the stronger mode. The amplitude of the standing wave for λ_2 is deliberately kept smaller to emphasize its feebleness. It should be noted that the weaker mode is not required to compete with the stronger mode for its survival as it basically feeds on the gain unseen by the other. In a nutshell, in an FP cavity, where the formation of a standing wave pattern is inevitable, a homogeneously broadened laser can therefore lase on multiple cavity modes by exploiting the phenomenon of spatial hole burning. It would be intriguing in this context to examine the case of a ring cavity that is capable of supporting a traveling wave.

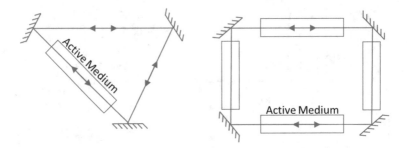

Fig. 8.10 Schematic representation of three- (left trace) and four-mirror (right trace) bidirectional ring cavities. The four-mirror ring is shown to contain active medium in all its arms basically to underline the fact that a ring cavity has a built-in ability to accommodate gain medium of longer length

8.5   Homogeneously Broadened Ring Cavity Laser

Unlike an FP cavity that is basically straight or linear, a ring cavity is not straight and is formed by at least three mirrors or more. Examples of ring cavities comprising three and four mirrors, which are most common, are depicted in Fig. 8.10. The active medium can be placed in any one of the arms of the cavity, as is shown in the case of the three-mirror ring cavity. The ring configuration, however, allows boosting the laser power by placing multiple gain cells in its different arms, as depicted in this figure. When the mirrors forming the cavity are perfectly aligned, the light will be amplified as it goes round and round inside it. A simple ring cavity, such as the ones shown here, will allow waves to travel in both forward and backward directions. The counterpropagating waves thus will once again interfere, as in the case of a linear FP cavity, with each other to produce a standing wave pattern. The phenomenon of spatial hole burning will therefore also be present here, and consequently, a homogeneously broadened laser will also lase here, like an FP cavity, on multiple modes, the only difference being the forward and backward waves will result in bidirectional emission from the ring laser. In the case of a four-mirror ring cavity, the two output beams will emerge orthogonally through the partially transparent mirror, as shown in Fig. 8.11. Understandably, therefore, to capitalize on the advantage offered by the ring cavity, one of the counterpropagating waves must be repressed. The ring cavity then essentially becomes a traveling wave cavity generating a single output beam. An optical valve that lets the passage of, say, the clockwise light and blocks the counterclockwise beam, much the same way as a diode blocks flow of electric current in one direction, will essentially effect operation of a unidirectional ring cavity (Fig. 8.12). An example of such an optical valve is a Faraday isolator, a device that rotates the polarization state of light by exploiting the magneto-optic effect[5] which is both expensive and delicate. A rudimentary optical apparatus that is capable

---

[5]The magneto-optic effect refers to the modulation of the properties of light when it interacts with magnetic field.

Fig. 8.11 A laser based on a simple ring cavity that intrinsically allows oscillation of both clockwise and counterclockwise waves will yield two output beams. In case of a four-mirror cavity, the two beams will emerge orthogonally

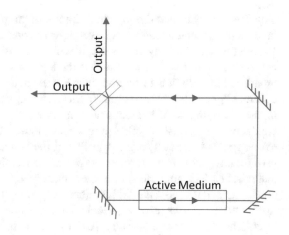

Fig. 8.12 Insertion of an optical valve, which blocks light moving in one direction and allows its passage in the opposite direction, will essentially make it a travelling wave cavity

Fig. 8.13 A telescope is an optical apparatus that would converge a beam of light in one direction and diverge it in the opposite direction

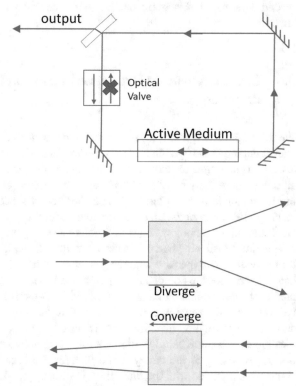

of providing selective loss to light in one direction can also prevent formation of standing waves when introduced inside the ring. A simple telescope that diverges light in one direction and converges in the other will also fit the bill as an unsophisticated optical valve here. The same is illustrated in Fig. 8.13. The converging light, by virtue of possessing a higher density of photons, will grow so rapidly by bleaching the gain that the diverging light will simply disappear in practically no time. In such a unidirectional or traveling wave ring cavity, where the formation of standing waves is clearly unfeasible, a homogeneously broadened laser will always lase on only one cavity mode that has the highest spectral gain. A homogeneously broadened unidirectional ring laser thus has the built-in capability of manufacturing coherent light with color of extreme purity. Although purity of color is compromised in the operation of a bidirectional ring laser, it, however, presents a situation conducive for a host of interesting applications. Most notable is the ring laser gyroscope that operates on the principle of the Sagnac effect [27] which is widely used for navigation in moving vessels like airplanes, ships, submarines, automobiles, and missiles. The ginormous ring lasers built specifically for LIGO (Laser Interferometer Gravitational-wave Observatory) experiments designed to detect gravitational waves are yet another widely known application of such a bidirectional laser. The laser gyroscope and LIGO system will be elaborated in detail in the second volume of this book dealing with the impact of lasers in science and humanity.

8.6   Inhomogeneously Broadened Laser and Spectral Hole Burning

We have up till now restricted ourselves to the manifestation of the phenomenon of gain clamping in the operation of only homogeneously broadened FP and ring cavity lasers. It is now time to examine as to how gain clamping will affect the operation of an inhomogeneously broadened laser. As the presence of inhomogeneous broadening is inevitable in a gas laser, we consider here a gaseous active medium placed inside an FP cavity of length l. The broadening of the gas laser, as a matter of fact, is an admixture of inhomogeneous and homogeneous components. Clearly the inhomogeneous broadening that originates from the velocity distribution of the gas atoms (or molecules) will increase with the gas temperature. The gas pressure, on the other hand, determines the homogeneous part of the broadening. For low or moderate gas pressures, the dominating mechanism of gain broadening will therefore be inhomogeneous, and the same applies to the present situation. With increasing pumping current, the population inversion and, in turn, the gain progressively increases. We pick a case when the gain has momentarily risen above the loss line setting off the process of lasing at the cavity mode that for the sake of convenience has been assumed to coincide with the line center frequency ν_0. The situation is qualitatively represented in Fig. 8.14. The group of atoms that can emit at ν_0 will only contribute to the process of lasing here as the gain stays clamped at this frequency. The gain at the frequencies neighboring to ν_0, being inhomogeneous, cannot participate in the process of lasing at this frequency and will remain unburnt. Consequently, a spectral hole will be burnt on the gain symmetrically about ν_0, and the width of the hole will understandably be homogeneous, governed basically by the gas pressure. As the

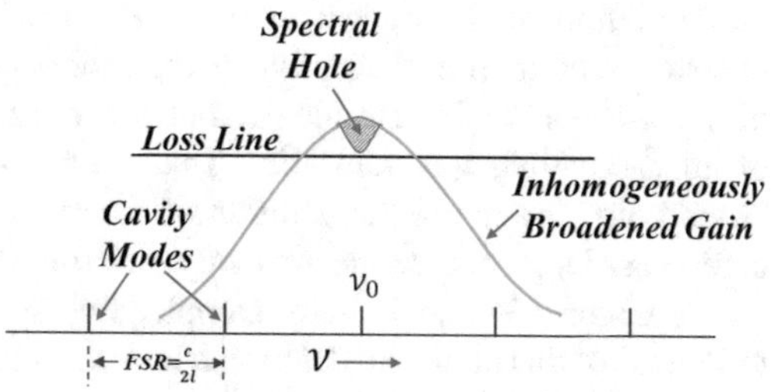

Fig. 8.14 Formation of spectral hole in the operation of an inhomogeneously broadened laser

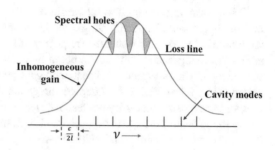

Fig. 8.15 Formation of spectral holes in the operation of an inhomogeneously broadened laser facilitates multimode operation of such lasers

FSR of the cavity in this example exceeds the lasing bandwidth, the lasing will be restricted only to the central mode that possesses the highest spectral gain. It is imperative at this point to examine the possibility of multimode lasing by considering a situation where the lasing bandwidth exceeds the FSR, and the same has been qualitatively illustrated in Fig. 8.15. Three cavity modes can be seen in this example to lie within the window of lasing. As, unlike in the case of homogeneous broadening, different groups of excited atoms are capable of emitting on these three different cavity mode frequencies, spectral holes will now burn on all three modes. This would facilitate the operation of the laser in multimode of which the line center mode possessing the highest spectral gain will be the dominant one. The spatial hole burning effect that is omnipresent in FP or bidirectional ring cavities will further aid the process of lasing on multiple modes just the way it happens for a homogeneously broadened laser. Removal of spatial effects by going for a unidirectional ring cavity cannot prevent lasing on multiple modes, as the prevailing spectral burning of holes in the gain will ensure their survival. Not surprisingly therefore, as far as purity of emitted light is concerned, an inhomogeneously broadened laser is prone to falling behind a laser with homogeneously broadened gain.

8.7   Spectral Hole Burning and Lamb Dip

Willis E. Lamb (1913–2008), regarded as a theoretician turned experimental physicists, predicted in 1964, a decade after winning Nobel Prize for discovering the famed Lamb shift, the occurrence of a dip at the line center in the output of a single mode Doppler broadened laser. Popular now as "Lamb dip," this is basically a manifestation of the optical Doppler effect. This dip occurring in lasers and its inverse counterpart in absorptive media enabled not only the accurate determination of the line center of an atomic (or molecular) transition but also the creation of light of mind-boggling spectral purity. That the unwrapping of physics enshrouding the "Lamb dip" has led to the addition of a new dimension to the science of spectroscopy, viz., Doppler free spectroscopy, offering unprecedented spectral resolution, is now a history. We shall study Doppler free spectroscopy in greater detail later in volume II of this book and determine how the advent of it has made possible to take a look where no one has looked before.

Toward gaining insight into the origin of the "Lamb dip," we take a closer look into the operation of a tunable single mode Doppler broadened laser. Tunability in the present context implies that the frequency of the oscillating mode can be scanned across the lasing bandwidth. Let us begin by considering a gaseous active medium with Doppler (inhomogeneous) broadened gain that is placed inside an FP cavity of length l (Fig. 8.16). Figure 8.17 depicts the gain as a function of frequency, the loss line, and a few pertinent cavity modes. In order to maintain a condition conducive for single mode lasing, the gain and loss are chosen here to ensure that the FSR of the cavity exceeds the lasing bandwidth. This will guarantee that at any point in time, only one mode can lie within the frequency domain $(\nu_B - \nu_A)$ of the gain intercepted by the loss line. Our endeavor here is to monitor the output power of the laser as a function of the frequency of the lasing mode by moving it across this lasing window from ν_A, the point of onset of lasing, to ν_B, the point of its termination. We know that the cavity mode frequency ν and the length l are related through the following cavity resonance equation:

$$\nu_n = \frac{nc}{2l}, \text{ where } n \text{ is the cavity mode index and } c \text{ is the velocity of light} \quad (8.4)$$

Fig. 8.16 Schematic of a single mode laser with provision of PZT driven cavity length tuning

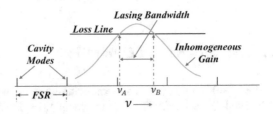

Fig. 8.17 This depicts the gain, the loss line, and a few cavity modes lying within and in the vicinity of the gain to essentially reflect the single mode character of this inhomogeneously broadened laser

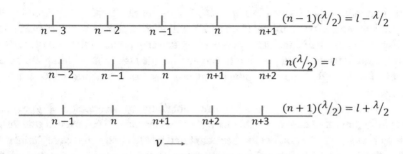

Fig. 8.18 This illustrates that the cavity mode belonging to a certain wavelength sweeps through the entire FSR if the cavity length is either increased or decreased by half that wavelength

Changing the cavity length is a way to shift the mode frequency, and this can be readily accomplished by translating any one of the mirrors forming the cavity. The obvious question that arises here is through what distance the mirror must be moved to ensure scanning of the entire lasing window by the lasing mode. To answer this question, we rewrite the resonance equation replacing frequency with wavelength as follows:

$$n \times \lambda/2 = l \tag{8.5}$$

$$\rightarrow n \times \lambda/2 \pm \frac{\lambda}{2} = l \pm \frac{\lambda}{2} \tag{8.6}$$

$$\rightarrow (n \pm 1)\frac{\lambda}{2} = l \pm \frac{\lambda}{2} \tag{8.7}$$

A comparison of Eqs. 8.7 and 8.5 helps us to conclude that if the cavity length is increased or decreased by half of the wavelength, then the mode belonging to this wavelength sweeps through the entire FSR of the cavity. This fact has been amply illustrated in the traces of Fig. 8.18. Thus, discerning the distance through which the mirror must move becomes straightforward once we know the fraction of FSR over which lasing persists. For instance, if the lasing bandwidth is half of the FSR, then any one of the cavity mirrors must move through $\lambda/4$ toward or away from the other. Considering the case of a visible laser that operates, say, at 500 nm wavelength, $\lambda/4$ measures to be 125 nm, a distance too small to be realized by the translation of a cavity mirror in a traditional manner. To this end, the piezoelectric transducers[6] (PZT), which exhibit extremely minute changes (to the tune of submicron level) in the longitudinal dimension upon application of a very high transverse electric field,

---

[6]A piezoelectric transducer (PZT) is a device that exhibits a minute change in length upon application of a voltage across its side. The cavity length of a laser can be controlled in a precise manner by mounting one of its cavity mirrors on a PZT.

come in handy. As illustrated in Fig. 8.16, one of the cavity mirrors, preferably the fully reflective one, when mounted on a PZT will allow a change in cavity length in a precise and controllable manner upon varying the magnitude of the voltage applied to it. We are thus now in a position to accomplish the task of monitoring the output power of this single mode Doppler broadened laser as a function of its emission frequency.

The condition prevailing here has been qualitatively described in Fig. 8.17. It essentially presents a case of single mode lasing as the cavity FSR exceeds the lasing bandwidth $\nu_B - \nu_A$. A situation has been captured where none of the cavity modes lie within the domain of lasing and the mode on which the onset of lasing is most likely to occur is located just at the left of ν_A. Upon application of an appropriate voltage across the PZT, this mode can be made to progressively move to the right. Lasing that sets in at the point of the mode crossing ν_A will continue until it moves past ν_B. The objective of this study is essentially to monitor the output of this single mode laser as the mode sweeps through the lasing bandwidth. Doppler broadening is inhomogeneous, and only a specific group of atoms contributes to lasing for each position of the cavity mode underneath the gain profile. The output power is thus expected to bear the same functional relationship with frequency as the gain. In reality, however, as shown in Fig. 8.19, the laser power exhibits a dip at the line center right where the gain actually peaks. The occurrence of such a line center dip on the power output of a single mode Doppler broadened laser was first theorized by Lamb in 1964 [28]. It is imperative at this point to gain physical insight into the origin of the Lamb dip that has a strong bearing on laser physics and applications.

We know from the optical Doppler effect, introduced earlier in this chapter (Sect. 8.2.2), that an excited atom, when at rest, emits light at the line center frequency and, when in motion, emits blueshifted light in the direction of its travel and redshifted light in the opposite direction. The emission from the single mode Doppler broadened laser vis-à-vis these three situations is illustrated in Fig. 8.20. Let us assume that the FP cavity of length l lies along the X-axis. In the active gaseous medium, the atoms move randomly in all possible directions and thus follow a Maxwellian velocity distribution. We, however, need to consider only the component of their velocity in the X direction, v_x, as the intracavity light travels, to and fro, along this direction alone. Let us first consider the case when the laser is tuned to ν_+, slightly

Fig. 8.19 As the frequency of a single mode gas laser is scanned across the Doppler broadened gain profile, its output shows a dip at the line center. The laser power attains maximum value at frequencies ν_+ and ν_- symmetrically located on either side of ν_0

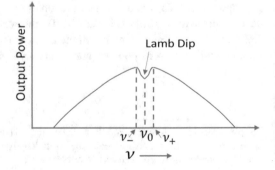

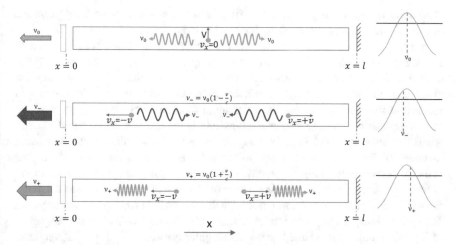

Fig. 8.20 Depending on the frequency at which the single mode laser has been tuned, different groups of atoms contribute to the process of lasing. Two velocity groups of atoms participate in lasing when the cavity mode is tuned above or below the line center, while only one group of atoms with zero velocity component along the resonator cavity contributes to lasing at the line center. Participation of two groups of atoms yields higher power, and the same is depicted in the lower and middle traces with wider arrows as the output. Participation of fewer atoms for lasing at the line center, on the other hand, results in occurrence of a dip in the power output here and consequently reduced output (indicated by a thinner arrow in the upper trace)

above the central frequency ν_0. This is therefore a case of a blue Doppler shift, meaning that the photons emitted along the direction of travel of the atoms will only be supported by the cavity. The group of excited atoms traveling with a longitudinal velocity v such that

$$\nu_+ = \nu_0\left(1 + \frac{v}{c}\right) \tag{8.8}$$

can only contribute to the output of the laser at this frequency. As shown in the lower trace of Fig. 8.20, there are basically two groups of atoms that participate here in the process of amplification of light through stimulated emissions; atoms moving with the velocity v toward the output mirror will be stimulated by light moving from right to left, and atoms moving with this velocity away from the output mirror will be stimulated by light going from left to right. Atoms moving with a velocity other than v cannot be stimulated to emit at ν_+.

In the second case, the laser has been tuned to ν_- slightly below the central frequency ν_0. This thus represents a case of red Doppler shift meaning that the photons emitted opposite to the direction of travel of the atoms will only be supported by the cavity (middle trace, Fig. 8.20). As in the previous case, there will again be two groups of atoms that contribute to the process of lasing here. However, the atoms moving toward the output mirror will now be stimulated by intracavity light traveling from left to right, and the atoms moving away from the output mirror will be stimulated by light going from right to left.

Finally, the laser is tuned to ν_0, the line center of the transition (top trace of Fig. 8.20). The atoms that can contribute to lasing at this frequency are obviously the ones that do not have any component of velocity along the X-axis, i.e., they are orthogonal to the laser cavity. Intracavity light circulating in both directions will succeed in stimulating only one group of atoms to yield output at ν_0. Thus, there are now fewer atoms available to manufacture light at the line center compared to cases when the laser is tuned slightly away from ν_0 on either side. With increasing detuning of the cavity from ν_0 on either side, atoms with progressively higher velocity components along the X-axis contribute to the process of lasing. As the number of atoms, with increasing velocity components along the laser cavity, monotonically reduces, the laser power will eventually drop with detuning after showing an initial rise in the vicinity of ν_0 on either side. Thus, when the frequency of a single mode laser is tuned across its Doppler broadened gain profile, the laser output will exhibit a dip at the line center. Clearly, this dip owes its origin to the ability of the two counterpropagating waves, which exist both in an FP and bidirectional ring cavity, to interact with atoms of two different velocity groups, under the detuned condition. It is not surprising that a unidirectional single mode ring laser will not exhibit the occurrence of a Lamb dip.

8.8   Lamb Dip and Frequency Stabilization of Gas Lasers

We have seen before (Sect. 6.6) that the fluctuation of the cavity length, inescapable in the operation of a laser, can easily mar the extreme spectral purity of the laser emission that basically stems from the ever-present spectral narrowing effect. In many applications of a laser, stability of emission frequency is an essential prerequisite. To develop an understanding of the dependence of the cavity length of a single mode laser on its frequency stability, we revisit the resonance equation of a cavity, namely,

$$\nu = \frac{nc}{2l} \tag{8.9}$$

where ν is the frequency of the nth mode of the laser cavity of length l and c is the velocity of light. In order to find out as to how the change in cavity length affects the frequency of the oscillating mode, we arrive at the following equation upon deriving Eq. 8.9:

$$d\nu = -\frac{nc}{2l^2} dl$$

$$\rightarrow \frac{d\nu}{\nu} = -\frac{dl}{l} \tag{8.10}$$

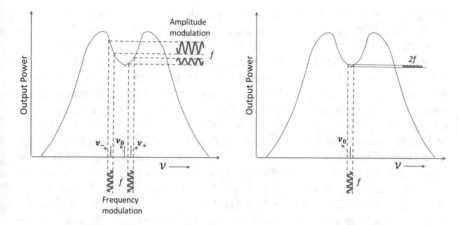

Fig. 8.21 Principle of "Lamb dip"-based frequency stabilization of a single mode Doppler broadened laser. The amplitude modulation of the laser output when the mode is tuned above and below ν_0 arising from the imposed frequency modulation is exactly out of phase (left trace). When the cavity mode is tuned to the line center, amplitude modulation vanishes at f and a weak modulation appears at 2f instead (right trace)

Clearly, any change in the cavity length is reflected in the frequency instability of the laser. Consequently, minimizing the fluctuation in the cavity length of the laser is the key to stabilizing its frequency. Although, as we have seen, the cavity length can be influenced by a number of factors, two common causes that affect it most detrimentally are temperature fluctuations and ground borne vibrations. Mounting the cavity optics on spacers made of invar[7] and the use of vibration dampers have been proven to be quite effective for stabilizing the laser cavity in a passive manner. Although some degree of stability is achievable through passive stabilization, it cannot safeguard against long-term instability. It is desirable to develop an alternative more rugged and reliable technique that would offer improved frequency stability in a persistent manner. As we shall find below, the occurrence of the line center dip on the output power of a single mode laser presents an ideal condition to develop precisely such a scheme of frequency stabilization.

The underlying principle operative here can be readily understood by referring to the illustrations of Fig. 8.21. As the cavity length of the single mode laser depicted in Fig. 8.16 is piezoelectrically tuned, making the mode frequency ν traverse across the lasing bandwidth, the generated power tuning curve will, as we know now, exhibit a dip at the line center, ν_0. We now impose a low frequency (f) sinusoidal modulation '$a \sin ft$' on the cavity length l such that $a \gg l$ and examine its impact on the laser

---

[7]Invar has very low thermal expansion coefficient, and therefore usage of invar spacers will make cavity length less sensitive to the temperature fluctuations.

power as a function of ν. Two situations, namely, when the cavity mode is tuned above ν_+ and below ν_- the line center, are illustrated in the left trace of Fig. 8.21. As seen, the modulation of power, which also occurs at f, is in phase with the length modulation when the cavity mode is located above ν_0 and out of phase when located below ν_0. However, when the cavity mode is tuned toward the line center from either side, the steady flattening of the curve manifests in a gradual reduction of the intensity of the power modulation dropping eventually to zero at ν_0. As seen on the right trace of this figure, while the modulation on power disappears at f at the line center, a weak modulation instead develops here at $2f$. The magnitude of the modulation of laser power therefore conveys information on the extent of detuning of the laser frequency from the line center, while its phase relative to the applied modulation indicates to which side of the dip the mode lies. The bottom line of "Lamb dip stabilization" is clearly therefore to monitor the modulation depth of the laser output and compare its phase with the imposed modulation on the cavity length all the while. An electronic servocontrol system, which forms the crux of this stabilization scheme, is shown in Fig. 8.22. The servomechanism essentially comprises an oscillator to impose PZT driven cavity length modulation, a photodetector to probe the modulated output, and a phase sensitive detection (PSD) system to compare the phases of input and output modulation and generation of the correction signal. Every time the cavity mode is drifted away from the line center, the PSD will generate a correction signal of appropriate magnitude and polarity that upon application to the PZT will pull the mode back to the line center. In the illustration of Fig. 8.22, both mirrors of the laser are PZT driven, and while modulation is imposed on the PZT holding the output mirror, the correction signal is fed to the other PZT.

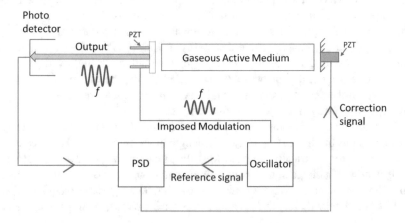

Fig. 8.22 A servomechanism upon integration with the piezo driven laser cavity generates the correction signal for application to the PZT to ensure locking of the lasing cavity mode at the line center

It is also possible to integrate the servocontrol mechanism into a laser driven by a single PZT. The popularity of Lamb dip frequency stabilization owes primarily to its inherent simplicity, and no wonder it is at the heart of a host of national standard laboratories spread across the globe like the erstwhile National Bureau of Standard (renamed National Institute of Standard Technology), USA; National Physical Laboratory, UK; and National Metrology Institute, Germany.

8.9   Inverse Lamb Dip and Stabilizing a Laser Away from the Line Center

Although "Lamb dip" offers extreme stability in the emission frequency of a laser, it however suffers from a major limitation as stabilization is possible only at the central frequency of the lasing transition. This drawback can be overcome by adding to the laser cavity, as shown in Fig. 8.23, a second cell containing another gas at very low pressure with an absorption frequency (ν_a) falling within the lasing bandwidth. This arrangement presents the possibility of using the absorption center of this gas as the new reference frequency in place of the line center frequency of the lasing transition. As we shall see below, this gaseous medium offers a ready prospect of stabilizing the laser emission on a frequency other than the laser line center. Just as it happens at the line center of the active medium, in the absorber, the two counterpropagating waves in the FP cavity will also undergo absorption by only one velocity group of atoms at the center of absorption. This will result in a narrow dip at the center of the absorption ν_a as depicted in the middle trace of Fig. 8.24. The spectral gain of the laser along with the lasing bandwidth is shown in the top trace. The convolution of

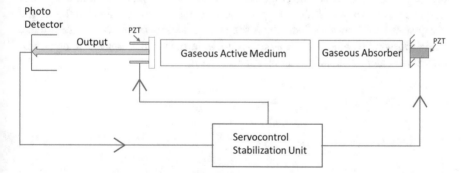

Fig. 8.23 Schematic of an "inverse Lamb dip"-based laser frequency stabilization setup. Usage of a gaseous absorbing medium in tandem with the gaseous active medium allows tuning of the cavity mode away from the line center for stabilization

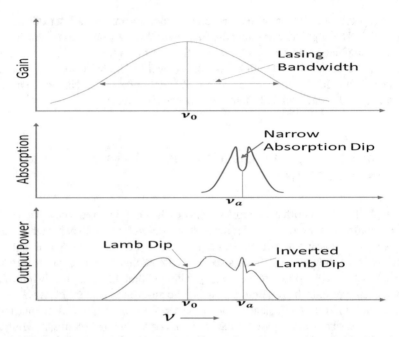

Fig. 8.24 Amalgamation of the gain in the active medium (upper trace) and absorption in the absorbing cell (middle trace) result in the occurrence of the inverted Lamb dip on the output power as the cavity mode is scanned across the lasing bandwidth of the single mode laser (bottom trace)

the two curves will manifest to produce a narrow "inverted Lamb dip" as the frequency of the single mode gas laser is tuned across its Doppler broadened gain profile (bottom trace). The inverted Lamb dip can be exploited to stabilize the laser frequency at ν_a, the absorption center of the gaseous species contained in the second cell just as the "Lamb dip" stabilizes the laser at its line center ν_0. It also offers a straightforward technique of locating ν_a and, in turn, provides a means to determine the exact center frequency of an optical transition.

Chapter 9
Boosting the Performance of a Pulsed Laser

9.1 Introduction

There are seemingly endless applications of lasers, such as medical, military, material processing, remote sensing, laser fusion, science with ultrahigh intensity light, etc., that call for coherent light of the kind of intensity a cw laser is incapable of delivering. As we have seen in Chap. 7, pulsing the laser can allow scaling up of the output power or intensity of a laser albeit over a limited period of time. Although specific applications will form the kernel of volume II of this book, we need to emphasize here that the requirement on the magnitude of intensity varies widely from application to application. For instance, while modest intensity over a relatively longer period may suffice for certain material processing applications, fundamental research aimed at unraveling the design of nature essentially calls for a flash of light of unthinkably high intensity. The very onset of stimulated emission, which is the key to the realization of a laser, restricts the peak power and, in turn, the intensity of the pulse it gives out. Practical limitation on the rate at which energy can be pumped into the active medium is yet another factor that also puts a brake on the maximum achievable peak power from a pulsed laser. These seemingly unsurmountable challenges have largely been overcome by compressing the duration of the pulse rather than enhancing its energy content.[1] The great advances made in the area of laser physics, not long after the first laser flashed its light, led to the conceptualization of techniques such as Q-switching, cavity dumping, and modelocking to achieve pulse compression. The practical implementation of these techniques was made possible by exploiting the remarkable advantages offered by saturable absorbers or devices such as electro-optic, acousto-optic, or optical Kerr media. From the modest

---

[1] Power, as we know, is the rate at which energy is delivered. Clearly therefore an ultrahigh power laser will be capable of delivering either enormously large energy over a modest length of time or a modest amount of energy over an extremely short duration.

D. J. Biswas, *A Beginner's Guide to Lasers and Their Applications, Part 1*,
Undergraduate Lecture Notes in Physics,
https://doi.org/10.1007/978-3-031-24330-1_9

189

beginning of the production of nanosecond optical pulses in the 1960s, we have now graduated through picosecond into the regime of femtosecond and even beyond. A femtosecond pulse straight out of the laser oscillator can accomplish seemingly unthinkable tasks. To name but a few, when focused on a tiny spot, it performs as a scalpel of extraordinary sharpness allowing a surgeon the luxury of performing a high-precision job such as a delicate eye surgery or drilling a tiny hole on the wall of the heart; tracking the minute details of all the intermediate species formed during a chemical reaction defeating the lightning speed of their formation, a concept that impacted the field of reaction dynamics so profoundly that its originator Ahmed Zewail (1946–2016), nicknamed the "father of femto-chemistry," was awarded the 1999 Nobel Prize in Chemistry; offering seemingly limitless bandwidth in data communication; etc. The energy content of this femtosecond pulse from the laser oscillator, however, needs to be elevated leaps and bounds before it reaches a level when, upon focusing, it would mimic the extreme condition that prevails at the center of a star, a pathway to big new science for knowing the unknown. As we shall see later in this chapter, the power of this ultrashort pulse can barely be boosted by letting it pass through an amplifier as its growing power invariably causes optical damage to the amplifier itself. Gerard Mourou (b–1944) and his doctoral student Donna Strickland (b–1959) at Michigan University succeeded in circumventing this problem in 1985 when they cleverly employed the chirped pulse amplification (CPA) technique. This made possible for the kilowatt (10^3) ultrashort (~100 fsec) oscillator pulse to leap coolly to tens of terawatt (10^{12}), a colossal optical power that even dwarfs the output of all the power plants of the world put together, although just at the moment of its flashing! It is no wonder that the conceptualization and implementation of the CPA fetched Mourou and Strickland a Nobel Prize in Physics, albeit a little belatedly in 2018. In addition to providing deeper physical insight into the various techniques to achieve pulse compression, this chapter will also address the challenges of amplifying an ultrashort pulse in the conventional way and overcoming them by adopting the technique of CPA.

9.2   Q-Switching

We know by now that the peak power of the laser pulse is essentially determined by the maximum value of the population inversion and, in turn, the gain that can be realized in the operation. In conventional operation, the phenomenon of gain clamping pins the gain down to the loss line and consequently puts a check on the maximum peak power of the emitted pulse. This situation is schematically illustrated in the top trace of Fig. 9.1. Clearly, the inability of the process of pumping to push the gain up beyond the loss causes the laser to underperform with respect to the emitted power. A possible remedy to this would be to push the loss line a long way up by way of appropriately augmenting the cavity loss. This would allow gain to rise much beyond what is possible in the conventional operation of the laser. An extreme

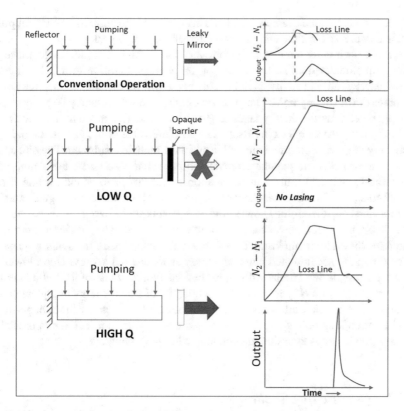

Fig. 9.1 In the conventional operation of a laser, when the gain can barely cross the loss, the rate of stimulated emission stays low and consequently burns up the energy stored in the population rather slowly. This, in turn, limits the peak power of the pulse (top trace). If a mirror of the cavity is blocked, the loss becomes much too big making room for the gain to build up to significantly higher value. No possibility of lasing here as the gain doesn't cross the loss line (middle trace). Before the energy stored as population inversion can appreciably leak through spontaneous emission, the mirror is unblocked restoring the cavity Q, and the extremely fast rate of stimulated emission rapidly brings out the stored energy as a giant pulse with greatly increased peak power

situation has been depicted in the middle trace of this figure where the cavity has been destroyed by blocking one of its mirrors with an opaque barrier. The loss that always overrides gain here prevents the onset of lasing, and the pumped energy stays locked up as population inversion, i.e., as the internal energy of the active medium. The possibility of loss of a part of this energy over time through spontaneous emission, however, cannot be ruled out. A lossy cavity essentially means a cavity with a low Q value. Thus, we might as well say that in a cavity with low Q, it is possible to establish a gain much beyond what is possible in a conventional cavity. Once the gain or population inversion has built up to the desired value, the trick would be to restore the cavity Q to its original high value by unblocking the mirror in a flash. The lasing now begins with a value of gain far exceeding the loss, and consequently, the profoundly increased rate of stimulated emission burns up the population inversion extremely rapidly. This results in the emission of a giant pulse that is compressed in time and possesses remarkably high peak power.

Clearly, switching the cavity Q from very low to high will initiate the pulse by setting up stimulated emissions to use up the inversion and, in turn, releasing the stored energy into the cavity as light. This pulse will be terminated once the cavity becomes devoid of any light. The duration of the pulse will basically be governed by the combined time the energy takes to get out of the active medium into the cavity and eventually out of the cavity itself. The absolute value of the gain at the point of switching of Q will determine the rapidity of the stimulated emission. The higher the gain is, the quicker the population inversion burns and vice versa. For a high gain laser, the Q-switched pulse can be as short as tens of nanoseconds, while it can extend to several hundred nanoseconds in the case of a relatively low gain laser.

In a nutshell, therefore, Q-switching is a technique to extract a short pulse of remarkably high peak power from a pulsed laser by modulating its intracavity loss. When the loss is high and lasing is prohibited, the active medium should be capable of retaining the accumulated pumped energy as its internal energy. Upon lowering the cavity loss, lasing begins with an exceptionally high value of gain allowing the active medium to quickly release the vast quantity of stored energy as a giant pulse. Clearly, therefore, not all lasers are Q-switchable. Only lasers with an upper laser level lifetime long enough so that the stored energy does not leak out appreciably through spontaneous emission are conducive to Q-switching.

9.2.1 Q-Switching Techniques

The Q-switch is essentially a device that should be able to switch the cavity loss from high to low quickly enough before the energy stored in the active medium leaks out substantially. Blocking and unblocking the laser cavity with an opaque barrier, considered earlier to elucidate the concept of Q-switching, may not be a very useful technique in reality; removing the barrier from the path of the oscillating light beam within the time light takes to complete one to and fro motion in a typical cavity seems quite impractical. The Q-switch devices, in operation today, can be broadly categorized into two types, active and passive. Active devices require power from an extraneous source for their operation and switching control, while a passive device is capable of operating on its own without any external help. The operation of various types of Q-switches is described in the following sections.

9.2.1.1 Active Q-Switching

The majority of the Q-switches that find applications are of an active nature. Some of the common active Q-switches include mechanical devices or optical modulators based on acousto-optic or electro-optic effects.

Fig. 9.2 Operation of a
rotating mirror Q-switched
laser. Lasing is impossible
whenever there is a cavity
mismatch (low Q, top trace).
Q-switched pulses emerge
every time one of the
reflective sides of the
rotating mirror gets aligned
with the other mirror
(high Q, bottom trace)

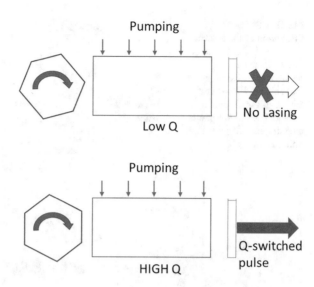

9.2.1.1.1   Mechanical Q-Switches

In the early days of lasers, mechanical devices, such as shutters, choppers, rotating
mirrors, or prisms, were used to effect the Q-switching operation. A mechanical
Q-switch with a hexagonal-shaped spinning mirror is shown in Fig. 9.2. The cavity
Q is switched from low to high every time a surface of the rotating mirror gets
aligned with the other mirror, albeit for a very short duration. This method, although
robust and suitable for use with high power lasers, suffers from low switching speed
and less reliability as it involves moving mechanical parts. It still finds applications,
limited nevertheless, primarily due to its operational simplicity.

9.2.1.1.2   Acousto-Optic Q-Switches

The acousto-optic (A-O) effect, as the name suggests, essentially addresses the
interaction of sound and light. It is well known that the sound wave, while traveling
through a medium, creates alternating regions of compression and rarefaction within
it in a periodic manner. In the optical context, these compressed and rarified regions
can be identified as zones of high and low refractive index (r.i.), respectively. Thus, a
sound wave passing through a medium causes periodic modulation of its r.i. along
the direction of motion of the wave. This essentially renders the medium as a
r.i. grating. When a beam of light passes through such a medium, this r.i. grating
acts like a series of slits and diffracts light out of the main beam in multiple directions
that manifest as different orders on a screen in the far field (Fig. 9.3). This means that

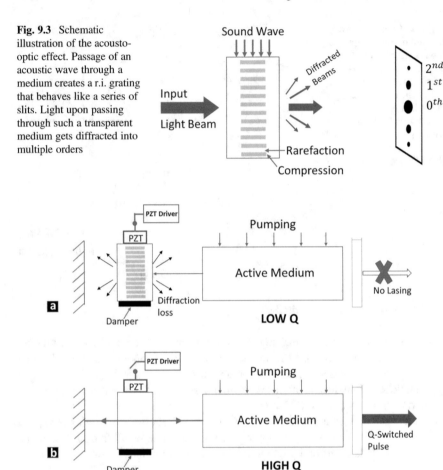

Fig. 9.3 Schematic illustration of the acousto-optic effect. Passage of an acoustic wave through a medium creates a r.i. grating that behaves like a series of slits. Light upon passing through such a transparent medium gets diffracted into multiple orders

Fig. 9.4 In order to effect Q-switching of a laser, the A-O modulator is placed between its active medium and one of the cavity mirrors. When the PZT is driven to send a sound wave into the modulator, it renders the cavity more lossy and prevents lasing (trace a). Disengagement of the PZT driver with the A-O device restores Q-value of the cavity prompting the laser to emit a giant pulse (trace b)

as the beam of light passes through such a transparent material, a part of it is spilled out as a result of the acousto-optic effect.[2]

The idea thus is to place an acousto-optic device into the laser cavity as shown in Fig. 9.4. As the gain begins to build, spontaneous emission also sets in. Application of an acoustic signal will initiate the acousto-optic effect resulting in the removal of a

---

[2]The diffraction of light into multiple orders as it interacts with an acoustic wave is called Raman-Nath effect [29]. There is yet another kind of acousto-optic modulator called the Bragg modulator [30] that diffracts light only in one direction. The Bragg modulator assumes importance when the light diffracted from the central beam constitutes the laser output as in the case of partial cavity dumping to be covered in a latter section of this chapter.

fraction of this spontaneous light from the intracavity photon flux and, in turn, lowering the cavity Q (Fig. 9.4a). As the acoustic signal is withdrawn and the acousto-optic effect disappears, the material will allow unhindered passage of light, switching the resonator Q to its original high value. It is pertinent here to discuss the feasibility of applying this concept to a practical laser system. For a visible laser, a transparent material such as quartz of cuboid geometry is often used as the acousto-optic device. The sound wave is created by applying a sinusoidal voltage to a piezoelectric transducer (PZT, introduced earlier in the preceding chapter, Section 8.7) and is launched into the quartz crystal by bonding the PZT to its one end (Fig. 9.4a). (In order to suppress the reflection of the acoustic wave from the opposite face of the crystal and rule out the possibility of its interference with the forward wave, a vibration damper is attached to its other end.) This allows gain to rise to a considerably higher value. Once the voltage applied to the PZT is withdrawn, it switches the cavity Q from low to high and the laser, in turn, gives out a giant pulse (Fig. 9.4b).

As the A-O modulator diffracts only a part of the intracavity light, it is therefore capable of introducing only a marginal loss. Not surprisingly therefore, an A-O Q-switch is not very effective in the case of a very high gain laser that by virtue of its high gain can overcome the small loss imposed by the A-O modulator and continue to lase. As will be seen later in this chapter, this disadvantage of the A-O Q-switch, however, can be exploited in its application as a partial cavity dumper. The A-O Q-switch also suffers from a low switching speed caused largely by the modest speed with which sound travels inside a material. Rapidity of the switching of Q from low to high basically depends on the time sound will take to get out of the path of the intracavity beam of light once the voltage to the PZT is withdrawn. For a typical beam diameter of ~1 cm and speed of sound in glass (~3000 m/s), the switching speed will work out to be several microseconds. Thus, the smaller the cross section of the laser beam is, the faster the switching speed.

9.2.1.1.3 Electro-Optic Effect and Birefringence vis-à-vis Q-Switching

Another class of switches that takes advantage of the electro-optic (E-O) effect has also emerged as a popular technique of Q-switching. This offers fast switching at the nanosecond timescale and is also capable of switching a high gain laser. It is imperative here to gain physical insight into the electro-optic effect before we discuss its exploitation to effect the Q-switching of a laser.

Certain materials exhibit a change in refractive index upon application of an electric field, and the magnitude of change is linearly[3] proportional to the applied field. This phenomenon was discovered by Friedrich Pockels (1865–1913) in 1906

---

[3] There is yet another effect called Kerr effect or quadratic electro-optic effect where the change in refractive index is proportional to the square of the electric field. Kerr effect is, however, generally much weaker than Pockels effect.

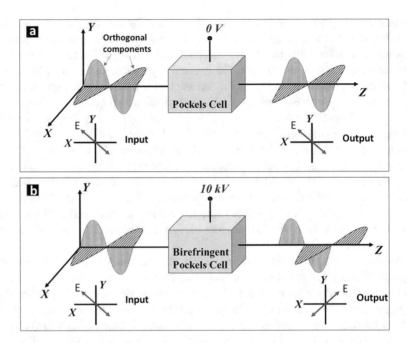

Fig. 9.5 When the Pockels cell is unbiased, there is no birefringence and light consequently passes through with unchanged polarization (a). When the Pockels cell is biased, the two components of polarization of the incident light travel at different speeds, and the light thus emerges with a rotated polarization

and is also known as the Pockels effect. We shall focus our attention here on those materials that, in addition to displaying the Pockels effect, also exhibit the phenomenon of birefringence[4] upon application of an electric field. The amalgamation of the E-O effect with birefringence leads to a remarkable prospect of achieving a controllable rotation of the polarization of light as it passes through a suitably biased Pockels cell of appropriate length. This fact has been illustrated in the traces of Fig. 9.5 that show propagation of light, linearly polarized on the X-Y plane, through both unbiased and biased Pockels cells. Having studied the "Behavior of Light" (Chap. 2, Section 2.8), we already know that the electric field of linearly polarized light can be split into two in-phase orthogonal components. When the Pockels cell is unbiased, it behaves like an ordinary transparent material through which both components travel at the same speed. The emerging light will therefore have the same polarization as the incoming light. However, when the Pockels cell is biased, the E-O effect sets in and alters the refractive index, and if the biasing also induces birefringence, the two in-phase polarization components of the incoming light will travel at different speeds. They will no longer be in phase anymore as one will fall

---

[4]Birefringence signifies a phenomenon where the refractive index of a material becomes dependent of the polarization of light passing through it. Such materials are called birefringent.

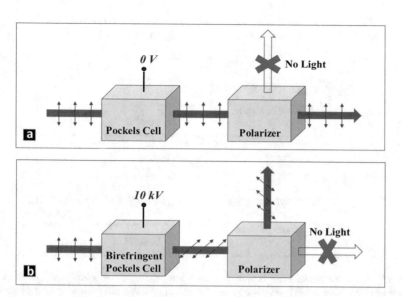

Fig. 9.6 Light passes unhindered through the polarizer when the Pockels cell is unbiased (trace a). When a biased Pockels cell rotates the polarization of light by 90° the Polarizer blocks the light (trace b)

behind or move ahead of the other. The extent of biasing and the length of the cell can be so adjusted as to make the phase of the two components differ exactly by $180°$. The vectorial addition of the electric fields of these two exactly out of phase components establishes beyond doubt that emerging light is now orthogonally polarized with respect to the light entering the Pockels cell.

9.2.1.1.3.1 Pockels Cell as a Light Switching Device

The ability of the Pockels cell to rotate the polarization of light by $90°$ is the key to its functioning as a device to switch light when used in conjunction with a polarizer[5] as illustrated in Fig. 9.6. When the Pockels cell is not biased, light travels through it with unrotated polarization, and a polarizer, placed next, transmits this light to the other side (Fig. 9.6a). However, when the Pockels cell is biased and passes the light by rotating its polarization through $90°$, the polarizer blocks this light completely (Fig. 9.6b). Clearly, a Pockels cell and a polarizer together are capable of destroying as well as restoring the Q-value of a resonator cavity and therefore, as described below, can be employed to Q-switch a laser.

---

[5]A polarizer, as we have seen earlier in Chap. 2, is an optical device that passes light on one polarization and blocks light on an orthogonal polarization.

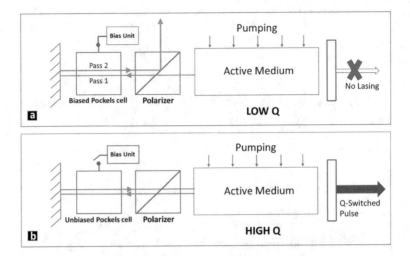

Fig. 9.7 When the Pockels cell is appropriately biased, the polarizer rejects any light from the active medium that reaches it after two passes through the Pockels cell preventing lasing (trace a). When the voltage to the Pockels cell is withdrawn, light passes through the polarizer unhindered, and cavity Q is restored to its original high value (trace b)

9.2.1.1.3.2   *Electro-Optic Q-Switches*

An electro-optic Q-switch that has two elements, namely, a Pockels cell and a polarizer, is placed between the active medium and one of the cavity mirrors as depicted in Fig. 9.7. Light from the gain medium passes through the Pockels cell twice before returning to the polarizer. For the polarizer to reject this returning light and prevent lasing, the Pockels cell should be so biased as to rotate the polarization by $45°$ in each pass (Fig. 9.7a). This allows the gain to rapidly build up until the biasing unit is switched off and the cavity Q is restored when the entire stored energy is emitted as a giant Q-switched pulse (Fig. 9.7b).

The general practice is to place the electro-optic Q-switch in front of the back mirror rather than the front mirror. With the back mirror being more reflective, a larger fraction of the light reaching here from the active medium is returned back to the polarizer boosting the performance of the Q-switch. The electro-optic Q-switch offers a very high switching speed as there is no moving part here as it happens to be in the case of a mechanical switch or a slow-moving sound wave for an acousto-optic switch. Furthermore, as the polarizer rejects the orthogonally polarized light in its entirety, this switch is capable of switching a high gain laser as well. The E-O switch is thus the most effective among all the active Q-switching devices, and no wonder it has emerged as the most popular Q-switching technique. However, electro-optic materials are not only very expensive but also require a high bias voltage for their operation. The rapid switching of such high voltages requires state of the art electronics.

9.2.1.2 Passive Q-Switching

The heart of passive Q-switching, also often referred to as self-switching, is a saturable absorber.

9.2.1.2.1 Working of a Saturable Absorber

It is imperative to gain insight into the working of a saturable absorber before we can appreciate its role in Q-switching a laser. To this end, we perform a simple experiment, the arrangement of which has been depicted in Fig. 9.8. The goal of the experiment is basically to monitor the transmission of light through a saturable absorber as a function of its intensity. The monochromatic light is derived from a laser, preferably a *cw* one, the frequency of which (ν) has an overlap with one of the absorption features (ν_0) of the absorber (inset of Fig. 9.8). The intensity of light transmitted by the absorber, I_{out}, as recorded by the detector is shown in Fig. 9.9 as a function of the intensity of incident light, I_{in}. In the beginning when the incident

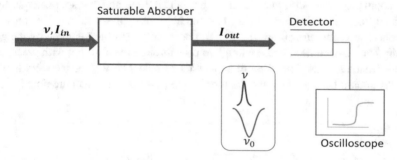

Fig. 9.8 A typical experimental arrangement to monitor the transmission of a saturable absorber as a function of the intensity of the incident light. The inset illustrates the overlap of the frequency ν of the incident light with the absorption center ν_0 of the absorber

Fig. 9.9 Beyond a threshold value of the incident intensity, the transmission by the absorber exhibits a steep rise

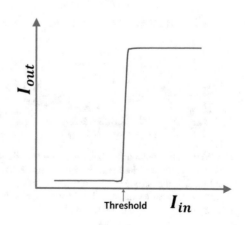

intensity is low, only a modest number of photons shine on the absorber and cannot escape absorption. With increasing intensity of the incident light, absorption also increases, and consequently the population of the corresponding excited level of the saturable absorber gradually swells. The absorber continues to prevent any leakage of the incident light to its other side until its intensity reaches a threshold value at which the excited level population equals that at the ground level. The obvious fall out of this is the saturation of absorption, and the medium's inability to remove any more photons from the incident beam of light results in a sudden rise of the output intensity beyond this point. (It is worthwhile to take another look at Fig. 4.5, Chap. 4, and the related text if you are hazy on this.)

9.2.1.2.2   Q-Switching with a Saturable Absorber

This remarkable ability of a saturable absorber to block light up to a certain intensity can be exploited to Q-switch a laser, and the arrangement of the resonator cavity to achieve this is shown in Fig. 9.10. The absorber is placed between the active medium and one of the cavity mirrors. In the beginning, when pumping has just commenced, the population inversion and, in turn, the gain is low, and the spontaneous light emitted from the active medium is weak and blocked by the absorber. This prevents the onset of lasing and allows the gain to rise rapidly, and light emitted by the gain medium soon becomes intense enough to pass through the absorber with practically no loss leading to the formation of a Q-switched pulse. Ideally the recovery time of the SA should be more than the duration of the pulse but less than the time interval

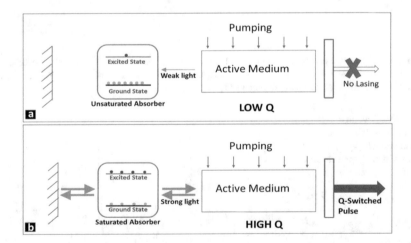

Fig. 9.10 In the beginning when the gain is low, the weak light from the active medium is blocked by the absorber and lasing is prevented (trace a). This allows gain to build up rapidly, and consequently the stronger spontaneously emitted light from the active medium saturates the absorption and passes through the absorber almost unhindered restoring the cavity Q and, in turn, forming the Q-switched pulse (trace b)

between two successive pulses. It is also desired that a SA be able to perform by absorbing only a small fraction of the energy of the generated pulse and thus will have a negligible effect on the power efficiency of the laser. Certain special dyes are known to perform satisfactorily as passive Q-switch devices. Although degradation of the dye over time is a major drawback, they find frequent applications primarily because of their inherent simplicity and cost-effectiveness.

9.3 Cavity Dumping

We know now that lasers with relatively short upper-level lifetime cannot be Q-switched. This is because leakage of the stored energy through spontaneous emission prevents any appreciable buildup of the laser gain, an essential prerequisite for effecting Q-switching. An alternative approach, called cavity dumping, can be employed for such lasers to achieve pulse compression. The short upper-level life may prevent the pumped energy from being stored as internal energy of the active medium, but it does not forbid its storage as radiation energy in the cavity. The scheme takes advantage of an extremely high Q cavity, with practically no leakage, to store photons emitted through stimulation. Once the number of intracavity photons has swelled to the required level, the cavity is destroyed, and these great many photons emerge in a flash forming a giant pulse of duration not exceeding the cavity round-trip time. The principle of operation can be understood by referring to Fig. 9.11, which depicts a rudimentary illustration of the experimental arrangement to effect cavity dumping. The cavity here is formed with two fully reflective mirrors. With practically no loss, the cavity Q, in sharp contrast to the case with Q-switching, is initially very high. Loss being very small, lasing begins almost in synchronism with the onset of pumping, and the generated light bounces back and forth between

Fig. 9.11 With the onset of pumping, stimulated emission begins to manufacture light that oscillate back and forth inside the cavity. When both the mirrors are fully reflective, the circulating light, with no escape route, rapidly builds up (a). Just at the instance of the termination of the pump pulse, a slight tilt of one of the mirrors will deflect all the light out in a flash as an extremely intense beam of light (b)

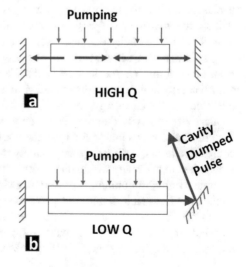

the mirrors and grows as long as the pump continues (Fig. 9.11a). The energy being dissipated by the pump is first stored briefly in the active medium as its internal energy before stimulated emission quickly realizes this as light. With no transmission loss, the intensity of light circulating inside the cavity can therefore reach an exceptionally high value by the time the pump pulse extinguishes. Destruction of the cavity at this point will result in the extraction of the entire accumulated light out of the cavity forming what can be termed as "cavity dumped pulse." Tilting of one of the cavity mirrors, as shown in Fig. 9.11b only for the purpose of comprehending the concept, will obviously reflect the beam out within a time that equals the cavity round-trip time of light. The first photon to emerge will be the one that happened to shine on the mirror just at the point of its tilting, while the last to emerge is the one reflected back into the cavity at that very instant. Obviously therefore, the duration of the cavity dumped pulse will be the time this last photon takes to complete one cavity round trip. Imparting a mechanical tilt to a mirror in a timescale commensurate with the rapidity of the pulse buildup seems quite impractical. An electro-optic switch, which we studied in-depth in the context of Q-switching, with the capability of transmitting or reflecting the light incident upon it, fits the bill quite well.

9.3.1 Cavity Dumping with Electro-Optic Switch

The experimental arrangement to employ an electro-optic switch to store electromagnetic energy in a high Q resonator cavity and then extract it in the form of an intense pulse by dumping the cavity is schematically illustrated in Fig. 9.12. When the Pockels cell is unbiased, the polarizer, as we already know, allows the passage of light in both directions. This makes possible to and fro lossless oscillation of light inside the cavity and, in turn, its rapid growth upon application of the pump pulse (Fig. 9.12a). In synchronism with the extinction of the pump pulse, the Pockels cell is biased, and consequently the polarizer passes light in one direction (pass 1) and reflects the return light (pass 2) out of the cavity producing the desired result (Fig. 9.12b). The cavity dumped laser, as a matter of fact, is just yet another kind of Q-switched laser where switching of cavity Q occurs in the reverse order, namely, from high to low Q. As the energy dissipated in the active medium is stored as optical energy in the cavity, the operation of a cavity dumped laser is independent of the lifetime of the energy levels involved in the process of lasing. Thus, the applicability of cavity dumping is not only limited to lasers with short upper-level lifetime that are unsuitable for Q-switching but also capable of driving Q-switch compatible lasers. As we have seen, cavity dumping allows extraction of the entire light stored in the cavity within one cavity round-trip time, and this is independent of the time required for building up the intracavity optical flux. The requirement placed on the switch is that it should be capable of switching in a time faster than the cavity round-trip time of the laser and the electro-optic devices offering nanosecond switching speed are thus compatible for cavity dumping operation. The time required by the pump source to supply energy into the active medium basically

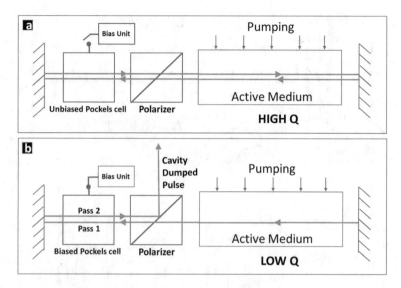

Fig. 9.12 The operation of a laser with an electro-optic cavity dumper. An unbiased Pockels cell allows to and fro oscillation and growth of circulating light (**a**). Biasing the Pockels cell lets the entire light escape through reflection at the polarizer forming an intense cavity dumped pulse of light (**b**)

governs the achievable repetition rate as the cavity dumped pulse is extremely short-lived. In contrast, a Q-switched pulse is much too long as it is essentially governed by the slow decay rate of the high Q cavity. This, in turn, limits the maximum achievable repetition rate of a Q-switched laser.

9.3.2   Partial Cavity Dumping with Acousto-Optic Switch

At the point of extraction of a cavity dumped pulse, the cavity is required to be prohibitively lossy, a condition that is generally met when an electro-optic switch is used. An acousto-optic switch, on the other hand, is capable of introducing a moderate loss to the cavity as it only deflects a fraction of the cavity light away and is therefore not a choice for effecting cavity dumping. However, in applications where only a part of the intracavity light is to be dumped out, the acousto-optic switch comes handy as a partial cavity dumper. Of particular interest in this context is the case of ion lasers that generally have a very short-lived upper level unsuitable for the storage of energy even for a reasonable length of time prerequisite for pulsed operation. Partial cavity dumping of such lasers when pumped in the *cw* mode can produce repetitive intense pulses. The typical experimental arrangement to partially cavity dump such a laser with an acousto-optic switch is shown in Fig. 9.13. The resonator cavity is understandably formed by two fully reflective mirrors, and the active medium is pumped continuously. Stimulated emission burns the gain, and the

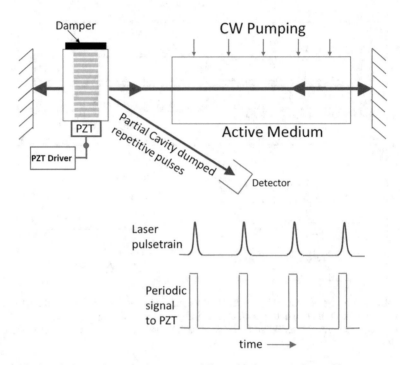

Fig. 9.13 A typical experimental scheme to partially cavity dump a cw laser with an acousto-optic dumper. A train of signal into the modulator and the corresponding laser pulses are also shown

intensity of intracavity light builds up rapidly with seemingly no escape route. Whenever a signal is impressed onto the PZT, the ensuing acousto-optic effect deflects a part of this circulating light as an intense pulse of coherent light. As this pulse is made up of only a fraction of the intracavity light and the energy is being pumped continuously into the active medium, it may not take much too long before the intracavity light is replenished, and the laser is ready for the next dump. Thus, upon partial cavity dumping, a laser, which otherwise is not conducive for repetitive operation, becomes high repetition rate compatible. A snapshot of the repetitive pulsing of the PZT and the corresponding laser pulse train is also shown in Fig. 9.13.

9.4   Modelocking

The Q-switched pulse, depending upon the typical laser parameters, can vary from tens of nanoseconds to hundreds of nanoseconds. The duration of a cavity dumped pulse, on the other hand, equals the time light takes to complete a round trip of the laser cavity and thus is much shorter and can be as small as a nanosecond for a moderately long cavity. A dramatic reduction in the pulse duration by up to three orders of magnitude is, however, possible when a multimode laser operates under

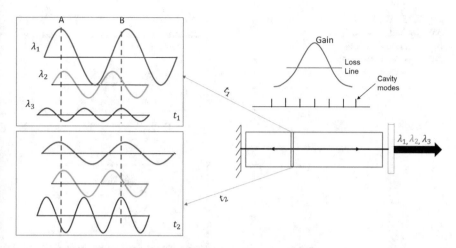

Fig. 9.14 Lasing is possible on three longitudinal modes of wavelengths λ_1, λ_2, and λ_3 for which the gain exceeds loss. A magnified view of these modes at an arbitrary location of the active medium is shown at two different times t_1 and t_2. For clarity, the modes are shown as spatially displaced in the transverse direction. For a free running laser, the location of the interfering peak changes with time

modelocked conditions. As we shall see below, the modelocking is required to be forced upon in the operation of a multimode laser that in the free running conditions operates with modes of randomly varying phases. Toward gaining physical insight into the phenomenon of modelocking, let us consider the operation of a laser wherein three longitudinal modes of wavelength λ_1, λ_2, and λ_3 ($\lambda_1 > \lambda_2 > \lambda_3$) are able to participate in lasing (Fig. 9.14). The electric field associated with all these three modes, which share the same gain volume, will interfere with each other to produce a resulting spatial intensity pattern. If the three modes are always in phase, then the profile of the spatial intensity pattern will remain invariant in time. In reality, however, phase of the oscillating modes of a free running laser will continuously vary in time and so would be the resulting spatial intensity pattern. This will manifest as irregular pulsation in the temporal emission of a multimode laser. To bring out this point, vital in the contest of modelocking, in a more revealing manner, we magnify an infinitesimally narrow strip of the gain medium and examine the interference pattern here at two instants of time, namely, t_1 and t_2. The magnified view of the three waves is spatially displaced in the illustration of Fig. 9.14 for clarity. As seen at time t_1, the three waves will interfere to produce a maximum field at the location A of the strip. As the phase of the waves varies randomly, at time t_2, the point of the maximum interference will shift to another location B. If the phase of the waves were to be somehow locked, then location A will continue to have the highest intensity. Considering now the entire length of the gain medium, there would be one location where the combined intensity of the three modes or equivalently the number of photons present will reach a maximum. If we ignore the tiny effect of dispersion, then the three waves would move at the same speed, and the photons in this packet

Fig. 9.15 The short modelocked pulse of light burns gain and rapidly grows as it moves along

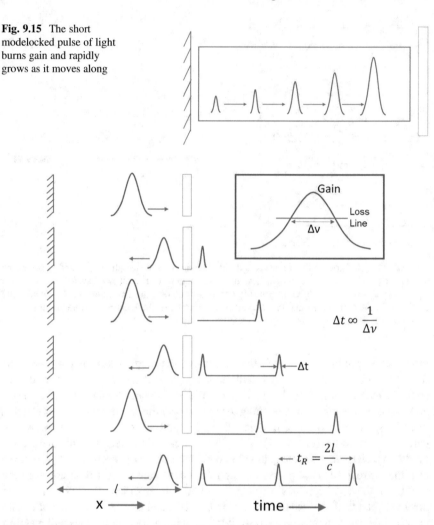

Fig. 9.16 The to and fro intracavity oscillation of the modelocked light pulse manifests in the emission of ultrafast train of pulses by the laser. Temporal width Δt of the pulses equals the inverse of lasing bandwidth Δν, while the periodicity of their emission is essentially the cavity round trip time t_R

too will move along at this speed and rapidly grow in numbers as the light circulates within the cavity (Fig. 9.15). Needless to say, we have considered here only the photons present at the location where interference yielded the highest intensity of light as they will outperform the photons present anywhere else in burning the gain through stimulated emission. Every time this bunch of intracavity photons bounces off the output mirror, a fraction of it leaks out and constitutes the modelocked train of pulses with a periodicity of cavity round-trip time, namely, $2l/c$ (Fig. 9.16). The duration of the pulse Δt is inversely proportional to the lasing bandwidth Δν. Thus, the modelocked pulses progressively shorten as an increasing number of modes participate in the lasing. It is therefore safe to say that the greater the broadening of the gain is, the higher the possibility of manufacturing shorter pulses through modelocking. For instance, while a Nd:YAG laser, with moderate broadening of gain, yields modelocked pulses of tens of picosecond duration, a Ti-sapphire laser, on the other hand, with enormous bandwidth can produce pulses well into the femtosecond regime.

It is straightforward to conclude now that while optical energy prevails across the entire length of the cavity in the Q-switching or cavity dumping cases, energy spreads only over a tiny fraction of the resonator length in the case of modelocking. A seemingly intuitive approach to effect modelocking is therefore to place a shutter into the cavity in the immediate vicinity of one of its mirrors that opens only for the desired pulse duration and remains shut for the rest of the cavity round-trip time. Although such a shutter is far from reality in a true sense, both active and passive switches, normally used to Q-switch a laser, can mimic such a shutter and facilitate the production of modelocked pulses. Like Q-switching devices, all the laser modelockers in operation today can be broadly categorized into two types, active and passive.

9.4.1 Active Modelocking

Active modelocking can be achieved with both electro-optic and acousto-optic switching devices. Here, we examine their functioning as modelockers.

9.4.1.1 Modelocking with Electro-Optic Switches

As we know, an unbiased electro-optic switch allows the passage of light and blocks it upon biasing. Clearly if this switch has to perform the role an optical shutter plays in the operation of a modelocked laser, the application and withdrawal of biasing voltage must be in sequence as shown in Fig. 9.17. Thus, the switch opens only once very briefly per round trip, which basically compacts the optical energy into a modelocked pulse with a spatial stretch shorter than the resonator cavity itself. The only light that can oscillate back and forth without any hindrance and burn the population inversion is the light that is packed inside this modelocked pulse. As illustrated in Fig. 9.16, this circulating pulse of light, which grows from short to shorter with the participation of an increasing number of modes in lasing, produces the modelocked train of pulses emitted by the laser.

Fig. 9.17 The sequence of biasing the Pockels cell that would make the electro-optic switch favorable to effect modelocking

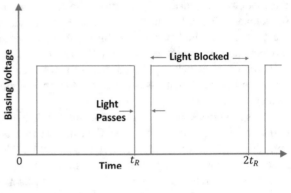

Fig. 9.18 An acousto-optic modelocker has a sound reflector at the rear

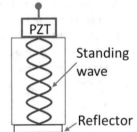

9.4.1.2   Modelocking with Acousto-Optic Switches

Unlike the electro-optic switch, which can completely block a beam of light, an acousto-optic device is able to slash only a fraction of light from the incident beam. Consequently, this device is unsuitable to perform as a modelocker in its conventional operation wherein the formation of an acoustic grating diffracts away a part of light from the incident beam. The trick here is to place a reflector at the other end of the transparent medium to send the transmitted acoustic wave back into the medium (Fig. 9.18). This, as we know, is in total contrast to the conventional operation of this device wherein a vibration damper is usually made use of to get rid of this transmitted sound wave. (It may be a good idea to refresh our memory on the working of an acousto-optic device by taking another look at Fig. 9.3 and the related text.) The reflected backward wave now interferes with the forward wave to create a standing wave akin to what is produced with a plucked string of a musical instrument. To understand the role standing waves play in the working of an acousto-optic modelocker, we take a closer look at the standing wave pattern formed in a string. The fundamental standing wave that has the longest wavelength is shown in the top trace of Fig. 9.19. Clearly the shape of the stretched string changes as it continues to oscillate between the two extremities. As seen, the string, bow shaped at the two extreme locations A and B, becomes perfectly straight in midway, represented as C, and this momentary state of straightness occurs twice in each cycle. Thus, it is prudent to believe that the string behaves as though it is not stretched at all twice per cycle. Drawing an analogy, we may now assume that the elastic medium of the

Fig. 9.19 This picture depicts a standing wave (top) followed by the shape of the string at a few specific instances during an entire cycle

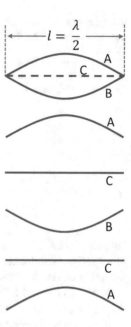

acousto-optic modulator that contains the stationary wave is also not perturbed twice per cycle. When the medium remains unaffected by the sound wave, the pulse of the modelocked light then passes through it unhindered without suffering any diffraction loss. Obviously if this device has to perform as a modelocker for a laser, then it must allow passage of a light pulse once per cavity round-trip time (t_R). If a signal of frequency f is applied to the PZT, then the standing wave developed in the medium will allow passage of light $2f$ number of times per second. This means that the modulator does not diffract light once every $1/2f$ second. The necessary condition for the synchronization of the arrival of the modelocked pulse at the modulator just at the instance of its opening is.

$$t_R = \frac{1}{2f}$$
$$\rightarrow \frac{2l}{c} = \frac{1}{2f},$$ where l and c are the cavity length and speed of light, respectively.
$$\rightarrow f = \frac{c}{4l}$$

A typical laser with a cavity length of *1.5 m* can thus be modelocked with a standing wave acousto-optic modelocker when a signal of frequency 50 MHz is applied to its transducer.

9.4.2   Passive Modelocking

This can generate pulses that are usually much shorter than what is possible with active modelocking. This technique is also more popular as the shutter performs without any extraneous intervention, has long operational life, and is not required to be serviced. A saturable absorber, the working of which is described earlier in this

Fig. 9.20 The recovery time of a saturable absorber Q-switch is too big compared to its switching time that is lesser than the cavity round trip time t_R

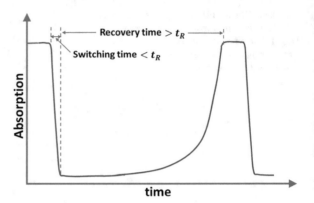

chapter (Sect. 9.2.1.2) and can passively Q-switch a laser, can also perform the role of a passive modelocker. To appreciate the performance of a saturable absorber as a modelocker, it is worthwhile to take a closer look at its functioning as a Q-switch, as illustrated in Fig. 9.20. The switching time (or alternatively the time for absorption to drop to a negligible value) must be smaller than the cavity round-trip time t_R to facilitate the rapid buildup of the Q-switched pulse. The recovery time of the switch, i.e., the time taken for the absorption to return to its original value, needs to exceed the duration of the Q-switched pulse lest the absorptive loss eats away a part of the pulse energy. A Q-switched pulse usually exceeds the cavity round time by many multiples; this essentially means that if a saturable absorber has to perform as a Q-switch, then its recovery time must far exceed t_R. In sharp contrast to Q-switching, modelocking, as we know, leads to the formation of a train of temporally sharp pulses separated from one another by t_R in time. Thus, for realization of modelocking, the saturable absorber must recover in a time quicker than t_R and be ready to be saturated again with the arrival of the next pulse of the circulating light. As we shall see below, an appropriately chosen saturable absorber can perform remarkably well as a modelocker for both linear and ring cavity lasers when it is located correctly inside the respective resonators.

9.4.2.1   Saturable Absorber: Modelocking of a Linear Cavity Laser

A schematic of saturable absorber assisted modelocking of a linear or Fabry-Perot cavity laser is shown in Fig. 9.21. The saturable absorber is placed in the immediate vicinity, often in optical contact, to one of the cavity mirrors. This is to ensure that the pulse of light can complete its to and fro journey comfortably through the absorber before its recovery. Although a saturable absorber is capable of modelocking both pulsed and *cw* lasers, we consider here only the case of *cw* lasers for ease of depiction and understanding. In order to examine the temporal evolution of the modelocked pulse, we consider the cases of both fast and slow absorbers as depicted in Fig. 9.22. A fast absorber can recover sufficiently quickly and almost as

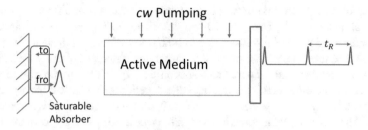

Fig. 9.21 In a linear cavity, the saturable absorber is placed close to one of the cavity mirrors, often the rear one, to ensure unhindered to and fro passage of the modelocked pulse through it

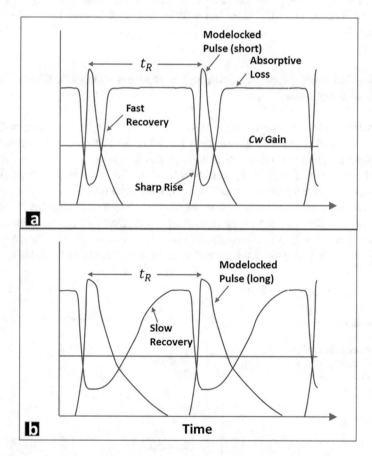

Fig. 9.22 The loss, gain, and power of optical pulse as a function of time for a fast (**a**) and slow (**b**) saturable absorber. The faster modulation of loss offered by a fast absorber makes the pulse shorter

fast as it takes to break into transmission, while the recovery of a slow absorber can extend almost up to the cavity round-trip time. The leading edge of the circulating light understandably suffers absorption in both cases which manifests in sharpening the pulse rise time. In the case of a fast absorber, the trailing edge of the pulse is also attenuated as it encounters appreciable absorption (Fig. 9.22a). In the case of a slow absorber, on the other hand, the trailing edge experiences much less absorption and consequently exhibits a gentler descent (Fig. 9.22b) and, in turn, tends to extend the pulse. The gain remains practically unaffected as the energy content of the modelocking pulses is too modest to burn it appreciably. In a nutshell, the fast absorber, by virtue of its ability to impose faster modulation of loss, tends to shorten the modelocked pulse. The decided advantage of colliding pulse modelocking, which is intrinsic in a ring laser and will be described below, essentially stems from this fact.

9.4.2.2 Saturable Absorber: Colliding Pulse Modelocking of a Ring Cavity Laser

One major hurdle encountered in modelocking a linear (also known as Fabry-Perot) cavity laser with a saturable absorber, in particular the fast one, is the problem in fabricating the absorber cell in optical contact with the cavity mirror. As we shall see now, the usage of a ring cavity readily overcomes this difficulty as it places no requirement of locating the absorber cell in the immediate proximity of cavity optics. The configuration of a saturable absorber assisted modelocked ring cavity laser is shown in Fig. 9.23. The ring cavity of length $4\,L$ comprises four mirrors with each of its arms being of length L. Although both the gain and absorption cells are shown here to be placed in one of the arms, the operation is possible by locating them

Fig. 9.23 A four mirror ring cavity containing the gain and absorption cells in one of its arms. The co- and counterpropagating pulses of the circulating light will always cross inside the absorber

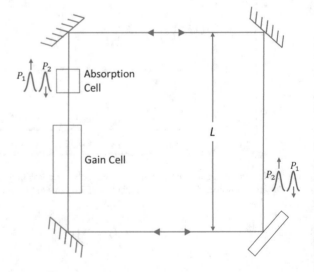

Fig. 9.24 When the gain
cell is spaced by quarter of
the cavity length from the
absorber, the pulses arrive
alternately in the gain cell
once per half of the cavity
round trip time

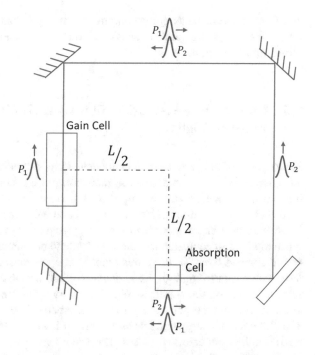

anywhere in the cavity. As we know in a ring cavity, two light pulses propagate in
clockwise and counterclockwise directions. The saturable absorber parameters are so
optimized that it would saturate only during the time the two pulses make an overlap
inside the absorber. The rest of the nonoverlapping light of both pulses will be
blocked by the absorber. The enhancement of light intensity at the instant of overlap
causes faster modulation of loss that manifests in the creation of two ultrafast pulses
P_1 and P_2 moving in opposite direction in a manner that they always cross inside the
absorber once per round trip.[6] The synchronized arrival of the pulses in the absorber
is automatic here and is independent of the location of the absorber with regard to the
gain cell. However, the appearance of these pulses in the gain cell depends on its
location with respect to the absorber. For instance, when the gain cell is located very
close to the absorber as shown in Fig. 9.23, P_2 clearly arrives at the gain cell almost
immediately upon the passage of P_1 through it. P_1, on the other hand, always reaches
the gain cell much longer after the passage of P_2 through it. This temporal asym-
metry in the time the two pulses take in reaching the gain cell after their respective
passage through the absorber can be overcome upon locating the absorber a quarter
of the cavity length away from the gain cell, and such a situation is depicted in
Fig. 9.24. As seen from this figure, the two pulses now arrive at the gain medium

---

[6]The interference of the two counterpropagating waves, although not explicitly mentioned here but
inevitable in a bidirectional ring cavity, results in manyfold increase of light intensity at the nodal
plane. This, in turn, results in even faster modulation of loss yielding better synchronization, higher
stability, and shorter pulses.

with the same time delay corresponding essentially to half of the cavity round-trip time. This allows the gain to always recover to the identical condition before the arrival of a pulse in it.

9.5 Chirped Pulse Amplification vis-à-vis Manufacturing Extreme Light

Exploitation of the phenomenon of modelocking, as we know now, has a decided advantage as far as generation of an ultrashort pulse is concerned. Furthermore, as the two successive pulses are temporally separated by the cavity round-trip time, a modelocked laser can thus give out pulses at an epically high repetition rate. This, however, comes at the expense of the energy contained in a pulse that essentially is much too modest compared to what is possible to be extracted from a Q-switched or cavity dumped laser. To drive this point home, we consider a typical modelocked laser with a cavity length of *1.5 m* delivering 100 watt average output power in the form of a train of 100 *fs* pulses. With the cavity round-trip time of this laser being 10 *ns*, it gives out one pulse every 10 ns or alternately 100 million pulses in a second. The 100 joule of energy that this laser emits in 1 second will therefore be carried by all 100 million modelocked pulses. The energy share of a lone pulse is thus just a mere one millionth of a joule, a ridiculously tiny magnitude to serve any meaningful purpose. Fortunately, the story does not end here though. True, a millionth of a joule is indeed too small, but the fact that the pulse is compacted in a time that lasts over just a tenth of a millionth of a millionth second[7] makes its peak power swell to an incredible magnitude of ten million watts. It is no wonder that these ultrashort pulse trains, although energetically insignificant, are powerful enough to fire the imagination of both scientists and technocrats across the globe. This has led to the proliferation of numerous applications, over the years, distinct enough to profoundly impact our everyday life and will form an important part of volume II of this book. Furthermore, the built-in focusability of these superpowerful ultrashort pulses to a tiny spot allows ablation of even the hardest material surface with unprecedented precision and control. This has led to the realization of microsurgeries on extremely sensitive and vulnerable parts of the human body that are unconceivable by a conventional tool or heat source.

There is yet another rapidly emerging area of research aimed at studying matter at extreme physical conditions such as that which exists at the core of a star. Simulation of such a condition in the confines of a laboratory is possible by focusing a light pulse of power exceeding a trillion watts to a tiny spot. This means that the million-watt power of the so-called ultrafast pulses must be scaled up by a factor of at least another million before they qualify to recreate such extreme conditions. Raising the power of this pulse even to a modest level by passing it through a light amplifier

---

[7]Please note that 100 *fs* or 100x10^{-15} second is equivalent to 10^{-1}x10^{-6}x10^{-6} second.

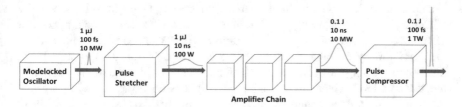

Fig. 9.25 Block diagram representation of a CPA cycle. The performance of the technique is exemplified here for a modelocked pulse of 1 μJ energy and 100 fs duration. The energy and duration of the modelocked pulse, however, can vary as long as the amplified power does not set off the detrimental nonlinear optical effect inside the amplifier

poses unsurmountable difficulties, let alone a millionfold boost. As the intensity of the pulse begins to grow inside the amplifier, unwanted nonlinear optical effects such as self-focusing[8] set in, impairing both the amplifying medium and the optics. It is worth mentioning in this context of another approach to develop a massive laser facility comprising a large number of independent oscillator-amplifier systems each capable of delivering a KJ of energy in pulses of nanosecond duration. The multiple outputs when combined into one raise the optical power beyond the terawatt level and are central to inertial confinement of fusion (ICF) research program. Notwithstanding the recent advances made toward ICF at the National Ignition Facility located at Livermore, California, this approach of raising power by building ever larger lasers did not gain popularity largely owing to their gigantic foot print and exorbitantly high cost. The seemingly insurmountable difficulty in realizing the passage of a femtosecond pulse through a light amplifier has eventually found a perfect solution in a magical concept now popular as chirped pulse amplification (CPA). The sequence of events that together represent the CPA is represented as a block diagram in Fig. 9.25. As seen, the femtosecond modelocked pulses are first stretched[9] by a factor that can be as high as 100,000. This operation understandably brings down the power of the pulse by the same factor. The standard light amplification technique can then be employed to raise its power. Clearly the power can be boosted by the same factor by which the pulse was initially stretched. Finally, a pulse compressor squeezes the pulse back to its original duration and, in turn, raises the power a hundred thousand times beyond the amplifier's capability. A typical

---

[8]Certain nonlinear materials, with intensity dependent r.i., behave like a lens and progressively focuses an intense beam of light with nonuniform spatial intensity profile, such as a Gaussian beam, passing through it. Self- focusing is a nonlinear optical phenomenon and will be described in detail in V-II of this book.

[9]A pulse stretcher basically is a dispersive element like a prism or grating that spreads the constituent colors of the femtosecond pulse spatially. Once the color components are spatially separated, they can be made to travel different distances and thus stretched in time. A compressor is again a dispersive element that negates the effect of stretcher and recombines the individual colors both in time and space. The literature is quite rich on CPA, and the interested readers may refer to the literature for developing a thorough understanding on the working of a pulse stretcher or compressor.

example will make this easier to understand. We once again begin with a seed pulse of a microjoule energy lasting over 100 femtoseconds. Upon stretching it by a factor of 100,000, the power of the pulse reduces from 10 MW to 100 W. Amplification of the pulse now by the same factor of 10^5 raises its energy from 1 μJ to 10^{-1} J, and finally compressing the pulse back to 100 femtoseconds will make its power swell to the terawatt level. Generally speaking, the light, with regard to its purity and intensity, has, to date, undergone two revolutions; first is of course the birth of laser in 1960, and the second is none other than the emergence of the CPA technique in 1985 to manufacture light of TW power or more from a table top laser. Indeed, after its birth in 1960, the implementation of CPA technique in 1985 brought about a youthful exuberance to the laser at 25, and, incidentally, even today when it is 60 plus, the exuberance is enduring.

Chapter 10
Different Types of Lasers

10.1 Introduction

The birth of lasers in 1960 has led to an unprecedented proliferation of research activity across the globe aimed at amplifying light in a wide variety of materials. Quite expectedly therefore, today we have a practically countless lasers emitting coherent radiation covering a significant part of the electromagnetic spectrum. There can be multiple ways of classifying a specific laser system. When they are categorized based on the physical state of their active medium, it is quite natural that they will be termed gas, liquid, or solid-state lasers. As we know, there exist a number of ways to pump a laser, and if they are arranged against the pumping mechanisms, it would be appropriate that they are classified as optically, electrically, chemically, gas dynamically, etc. pumped lasers. If they are organized based on the wavelength of their emission, they would most appropriately be termed as infrared, visible, ultraviolet, vacuum ultraviolet lasers, and so on. Here, we adopt the grouping based on the state of the active medium, as it allows a representative description of the large variety of lasers in an orderly and comprehensive manner in the present context. Other kinds of lasers, the workings of which are distinctly different and do not truly qualify to be grouped with the majority of conventional lasers but nevertheless stand pretty tall in the laser world, will be addressed individually. They include semiconductor lasers, fiber lasers, excimer lasers, chemical lasers, free electron lasers, and gas dynamic lasers.

© The Author(s), under exclusive license to Springer Nature Switzerland AG 2023 217
D. J. Biswas, *A Beginner's Guide to Lasers and Their Applications, Part 1*,
Undergraduate Lecture Notes in Physics,
https://doi.org/10.1007/978-3-031-24330-1_10

10.2   Gas Lasers

The atoms or molecules are not tightly packed in the gas as they are in the case of a liquid or solid. A gaseous medium thus presents a condition that is spectroscopically relatively simpler to study, and consequently extensive data on their energy levels existed even prior to 1960, the year the laser was born. We also know that striking an electric discharge is fairly straightforward in a gas. Furthermore, it is common knowledge that the excited energy state of a gaseous species can be readily populated through inelastic collisions with electrons present in the discharge in much the same way a fluorescent light source derives its energy. Understandably, in the mid-1950s when the laser concept was being developed following the path breaking works of Nobel Laureates Townes, Basov, and Prokhorov, a gaseous medium appeared to be the most attractive active medium for the operation of the very first laser. However, then as luck would have it, a flashlamp pumped solid-state medium earned the distinction of allowing amplification of light by stimulated emission of radiation for the first time ever. However, the gaseous medium was not very far away as the operation of the next laser that was reported in December 1960 was indeed a gas laser, namely, a *cw* He-Ne laser. Researchers over the years have literally discovered thousands of new lasing transitions in a wide variety of gases spanning across a broad range of electromagnetic spectra, although only a few have proven to be beneficial from the point of view of applications. Once indomitable in the field, the importance of gas lasers is being eclipsed of late by the growing technological superiority of semiconductor and solid-state lasers. Gas lasers, however, still at work today in certain areas primarily because they intrinsically offer superior spatial and spectral quality of the beam and provide higher power over some part of the spectrum. He-Ne lasers operating in the visible region are indispensable as alignment lasers even today. The role of CO_2 lasers operating in the infrared in industrial and medical applications is quite well known. The multiple applications of excimer lasers,[1] operating in UV and deep UV, such as the manufacturing of semiconductors and medical and scientific research, are also well recognized. The lasing species in a gas laser can be atoms, molecules, or ions, and consequently we shall choose the three most popular lasers from these three categories and describe their operations. However, unlike atomic and ionic gas lasers, where lasing transitions occur between electronic energy levels, the vibrations and rotations of constituent atoms in a molecule add diversity and richness to the emission from a molecular gas laser. The operation of these lasers will be described in a separate chapter (Chap. 11), as a basic knowledge of molecular spectroscopy is prerequisite to appreciate their working.

---

[1] Although the lasing medium is a gas here, this class of lasers will be described under a different heading as the way they work distinctly differs from the other kind of gas lasers.

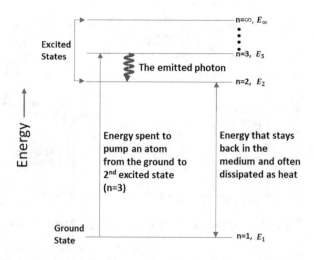

Fig. 10.1 Qualitative representation of the ground state and the first few excited states of a typical atom

10.2.1 Atomic Gas Lasers

We know from Chap. 4 that an electron when bound to an atom can possess only discrete values of energy. We also know that the magnitude of the energy of the electron in an excited state E_n is inversely proportional to the square of the orbital quantum number n of this state. It is appropriate to examine here the bearing of this inverse square dependence, namely, $E_n \propto \frac{1}{n^2}$, on the quantum efficiency of an atomic laser. The inverse square dependence essentially manifests as the electronic energy levels getting increasing closer to each other with increasing excitation. This situation has been qualitatively illustrated in Fig. 10.1, which shows the energy of the ground (E_1) and the first (E_2) and second (E_3) electronic excited states of a typical atom. The fact that $(E_3 - E_2) \gg (E_2 - E_1)$ is clearly revealed here albeit qualitatively. Let us now take a closer look at the operation of this atomic system as a laser with the second excited state of energy E_3 as the upper laser level and the first excited state of energy E_2 as the lower laser level. In the beginning, practically all the atoms reside at the ground state of energy E_1. In order to create population inversion, atoms from the ground state must be selectively transferred to the upper laser level. To move just a lone atom from the ground to the upper level, we need to expend an amount of energy of magnitude $(E_3 - E_1)$. A small fraction of this energy, namely, $(E_3 - E_2)$, will be carried by the photons originating from the emission between the upper and lower laser levels. The bulk of the absorbed energy, i.e., $(E_2 - E_1)$, stays back in the system and makes no contribution to the power output of the laser. The quantum efficiency of the laser, namely, $\frac{E_3 - E_2}{E_3 - E_1}$, is intrinsically small for an atomic system primarily because the lasing transition occurs far above the ground state.[2] In

---

[2]This also explains as to why atomic gas lasers predominantly emit in the visible and near infrared regions of the e-m spectrum.

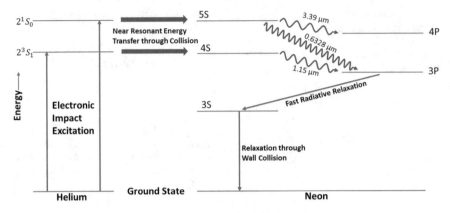

Fig. 10.2 Energy levels of helium and neon that participate in the process of lasing

practice, the working efficiency of an atomic gas laser will be considerably lower than the quantum efficiency. This is because the pump energy is realized as the kinetic energy of the electrons in the discharge following a Maxwellian distribution, and so a fraction of these electrons can only populate the upper laser level efficiently through collisions. As a matter of fact, the working efficiency of an atomic gas laser can typically lie within 0.01–0.1%.

10.2.1.1   Helium-Neon Lasers

The He-Ne laser has the distinction of not only the first gas laser but also the first *cw* laser. Ever since its invention in late 1960 by Iranian-American scientist Ali Javan (1926–2016) and coworkers at the Bell Telephone Laboratories, Murray Hills, NJ, USA, it has established itself as one of the most common and popular gas lasers. The laser operates by striking an electric discharge in a low-pressure gaseous mixture comprising five parts helium and one part neon. Neon plays the role of light emitting species, while helium helps create and maintain population inversion. The electronic energy levels of both helium and neon that participate in the process of lasing are shown in Fig. 10.2. The second orbit ($n = 2$) of the electron in the helium atom is split into two long-lived metastable sublevels[3] identified as 2^1S_0 and 2^3S_1. The inelastic collisions that the helium atoms in the electronic ground state undergo with the electrons present in the discharge result in their excitation to both these levels.

---

[3] The electron rotating in an orbit intrinsically possesses an angular momentum, and further as it also spins about itself (akin to Earth's rotation about the Sun and itself), it is also endowed a spin angular momentum as well. Like the electron's energy, its angular momentum is also quantized. These effects cause the energy levels of the electrons in the various excited states to split into sublevels denoted by the orbital quantum number n, angular quantum number l, and its spin s. The readers may please refer to any textbook on modern physics in this context.

$$He(\text{Ground state}) + e(\text{KE}) = He(2S_0) + e(\text{Less KE}) \tag{10.1}$$

and

$$He(\text{Ground state}) + e(\text{KE}) = He(2S_1) + e(\text{Less KE}) \tag{10.2}$$

These levels have a very close match in energy with the 5S and 4S levels of neon. Consequently, as the excited helium atoms collide with ground state neon atoms, near resonant exchange of energy takes place between $2S_0$ and 5S and between $2S_1$ and 4S.

$$He(2S_0) + Ne(\text{Ground state}) = Ne(5S) + He(\text{Ground state}) \tag{10.3}$$

and

$$He(2S_1) + Ne(\text{Ground state}) = Ne(4S) + He(\text{Ground state}) \tag{10.4}$$

The slight mismatch of energy is well compensated by the omnipresent thermal energy. The transfer of energy from helium to the energy states of neon creates population inversion on several transitions, the strongest being that at 1.15 μm. Consequently, the very first He-Ne laser operated on this infrared line. Any other transitions can be made to lase by providing selective losses to the other competing transitions, readily doable by proper selection of cavity mirrors. In the visible range, the strongest line is red (632.8 nm), while the orange (612 nm), yellow (594 nm), and green (543.5 nm) lines are very weak and understandably have no commercial value. The lower laser levels are energetically located much above the ground level, and collisions of neon atoms with the tube wall help dissipate this energy. To facilitate the relaxation of energy this way, narrow bore tubes are therefore preferred. Power scaling of the He-Ne lasers is thus not possible by increasing the cross section of the discharge tube. The extractable power from a typical He-Ne laser essentially depends on the lasing transition, and commercial lasers are available in the power range typically between a milliwatt to tens of milliwatts. Once the neon atoms return to the ground state, they become available for re-excitation and, in turn, continuation of the *cw* operation of the laser. As the operating pressure is low, the broadening of gain is predominantly Doppler with a full width of ~1.5 GHz. Thus, for a cavity with a typical length of 15–50 cm, several longitudinal modes can simultaneously lase. However, single longitudinal mode (SLM) operation is nevertheless possible by judiciously locating the loss line. For any special applications that require a large coherence length,[4] SLM He-Ne lasers are also commercially available. Some of these applications, such as holography, interferometry, and optical clocks, will be

---

[4]Coherence length of a laser is a measure of its spectral purity and mathematically can be expressed as the inverse of the spectral width of the laser emission. Coherence length and its bearing on the performance of a laser will be elaborated in detail in volume II of this book.

covered in the second volume of this book. The general applications of the He-Ne laser include its use as a reference beam for the precise alignment of critical components in the construction industry, laser resonator cavity, sewer pipe, etc.; as a tool to read the digitally encoded barcodes located on products in supermarkets; as educational aid in optics laboratories of colleges; and as research tools in academic institutes. One of its infrared emissions at 1.15 μm is employed for the measurement of optical fiber transmission lines that have a very insignificant loss at this wavelength. Gyroscopic applications of a ring cavity He-Ne laser are also quite well known.

10.2.2   Ion Lasers

In atomic gas lasers, the electrons involved in the lasing transitions are essentially the valence or outermost electrons and are capable of generating light across visible and near-infrared regions of the e-m spectrum. For the generation of light at the shorter end of the visible and into UV and even beyond, clearly the inner electrons, which are more strongly bound to the nucleus, are to be engaged in the process of lasing. This realization led to the first observation of laser action in gaseous ions by William Earl Bell[5] (1921–1991) in early 1964 [31], not long after the discovery of lasing in atomic and molecular gases. By the end of 1964, laser action was reported in the ions of 11 gases that resulted in the addition of over 200 more wavelengths to the rapidly growing list of new lasers at that time. Similar to most atomic and molecular gas lasers, ion lasers are also pumped by taking advantage of gas discharge. As energy needs to be expended to strip the atom of one or more electrons before causing its excitation, the ion lasers can convert only a tiny fraction of the pumped energy into the creation of coherent light. As the generation of lasing in deep UV requires doubly ionized ions, they are characterized by even poorer efficiency. It is not surprising that the wall-plug efficiency of ion lasers can be as low as 0.001% and typically lies within 0.01–0.001%. The operation of one of the most popular ion lasers, namely, the argon ion laser, is described below.

10.2.2.1   Argon Ion Lasers

The argon ion laser was first operated in 1964 by William Bridges (b-1934) at the Hughes Aircraft Laboratory [32] soon after the invention of this class of lasers. This is the most powerful *cw* laser in the visible and *UV* wavelengths and can produce tens of watts of output in conventional operation. The most popular wavelengths are

---

[5]The research work of William Bell, inventor of ion lasers, led to the development of the first atomic clock used in the Apollo missions to the Moon. The atomic clock he made is now on display along with his handwritten notes in the Smithsonian Museum of American History.

458 nm (blue), 488 nm (blue-green), 514 nm (green), and 351 nm (*UV*). As more heat is dissipated in an ion laser compared to other gas discharge-based lasers, a ceramic discharge tube is often used for the operation of an Ar ion laser. The basic construction features are qualitatively similar to other low-pressure gas discharge lasers except that a solenoid spirals around the discharge tube of all but low power *Ar* ion lasers. This electromagnet coil wound across the plasma tube generates a strong magnetic field longitudinally along its axis. This essentially increases the current density by keeping the electrons and ions near the center, rather than allowing them to collide with the walls of the discharge tube. The better confinement of plasma, in turn, allows scaling up of power and is particularly useful for *UV* emissions that have inherently low power conversion efficiency. The outer wall of the discharge tube of the low power (<100 mW) lasers is air-cooled, but with increasing power, it becomes far too hot and is required to be water-cooled often by incorporating a recirculating cooling system containing a chiller.

A part of the KE of the electrons, as we know, is expended in ionizing the atoms, and another part is spent in raising the ions to higher energy levels. There exist a number of metastable states to which the excited ion can drop down, and they serve as the upper level of several possible laser transitions. In the case of a nondispersive cavity, feedback is possible on lights emanating from all these transitions, and the laser, in turn, will emit on multiple wavelengths (Fig. 10.3a). Usage of an intracavity dispersive element such as a prism will allow cavity feedback only on one wavelength at a time resulting in single wavelength lasing (Fig. 10.3b). The dominant broadening is of Doppler origin and has a linewidth of approximately 5 GHz that would encompass many longitudinal modes for a laser of typical cavity lengths.

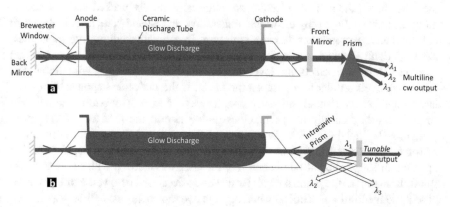

Fig. 10.3 In a nondispersive cavity, feedback and, in turn, lasing is possible on all the transitions for which round-trip gain exceeds the loss. Subsequently, a prism, kept outside the cavity, disperses the laser output into its constituent wavelengths (**a**). In a dispersive cavity, on the other hand, the dispersive element, such as a prism, disperses the intracavity light spatially allowing thereby feedback only on a single wavelength at one time. As seen here, the light at wavelength λ_1 falls on the output mirror normally and therefore receives feedback, while other wavelengths, like λ_2 and λ_3, either fall obliquely on this mirror or miss it completely and get lost (**b**)

Thus, a dispersive cavity laser will operate on a single line but on multiple longitudinal modes. The standard laser resonator is so chosen as to have a close match of the TEM_{00} mode and the diameter of the discharge plasma to produce a low divergent beam. High-power lasers, however, often oscillate on multiple transverse modes.

10.3   Liquid Lasers

In the early years of laser development, the researcher community's endeavor to find newer lasing transitions was centered primarily on gaseous and solid-state active media, while any major activity on liquid media was conspicuously absent. Following its invention in 1960, the laser inventory grew by leaps and bounds, and by 1965 there were lasers providing coherent light covering the infrared, visible, and near UV with potential applications in almost all branches of science and engineering. The only exception, however, was the field of spectroscopy; the inability of a gaseous or solid-state medium to provide a continuously tunable laser emission deprived the spectroscopists of reaping any major benefit of the unique light of a laser. The lasing transitions of atomic or ionic gas lasers are purely electronic in origin, and it is no wonder that they emit only at specific wavelengths. As seen in Sect. 11.2 (Chap. 11), in the case of a molecule, the constituent atoms can both vibrate about the CM and also rotate about the internuclear axis. Thus, a molecule, in addition to the electronic transition, also exhibits ro-vibrational transitions that, although are much closely spaced in wavelength, do not overlap at the typical gas pressures. This nevertheless allows a molecular gas laser to emit on discretely tunable wavelengths. In a solid, on the other hand, the atoms/molecules are extremely closely packed and not free to move and thus exhibit only electronic transitions, and a solid-state laser is therefore not conducive to rendering tunable emission. In a liquid medium, the molecules, although tightly packed, are flexible enough to exhibit both vibrational and rotational motions. Thus, each electronic level will be associated with many vibrational and rotational levels similar to a gaseous molecule. If the molecule happens to be very large, then these rotational and vibrational levels tend to be so enormous and closely spaced that they simply merge together giving rise to the prospect of providing continuously tunable light. Europium chelate, a metalloorganic compound, dissolved in a solvent served as the active medium of the very first liquid laser that operated in 1965. The pulsed operation of the laser was realized with a conventional spiral flashlamp as the pump source. However, as described below, the true potential of a liquid medium was realized when organic dye molecules made their appearance in 1966.

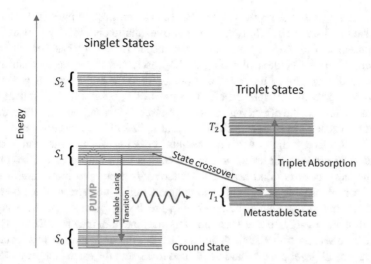

Fig. 10.4   A few low-lying electronic energy bands, both singlets and triplets, of the dye molecules that participate in the process of lasing

10.3.1   Dye Lasers

The dye laser was invented independently by Peter Sorokin (1931–2015), an American-Russian physicist, and Fritz Peter Schäfer (1931–2011), a German physicist, in 1966. Liquid solutions of organic dye compounds performed the role of the active medium in these efforts. Incidentally, a ruby laser was used as the pump source, and near-infrared dye was used as the lasing medium in both of these works. Consideration of the energy levels that participate in the process of lasing, as shown in Fig. 10.4, will help understand the working of a dye laser. The two types of electronic states that a molecule can have, viz., singlet and triplet states,[6] have been distinguished here as they play totally contrasting roles in the operation of the laser. Being a liquid medium, as we know now, these levels will be greatly broadened in energy as they are composed of numerous rotational and vibrational sublevels.

In thermal equilibrium, majority of the population resides at and around the bottom of the singlet ground state S_0. The pump wavelength is so chosen as to interact with this group of population and transfer them close to the top of the first excited singlet state S_1 to derive the maximum benefit in terms of both tunability and operating efficiency of the laser. Once excited, the molecules begin to quickly relax down the vibrational levels within S_1. However, if the dye cell is placed inside an

---

[6]The singlet and triplet states of a system are categorized depending on the orientation of the spins of the electrons that make up the system. When spins of all the electrons are paired, the resulting quantum state is a singlet state, while the case of electrons with unpaired spins is characterized as a triplet state. In the case of a molecule, the ground state is usually a singlet state, while the excited state involving an excited electron can be both singlet and triplet.

optical cavity, then before the excited molecules can decay appreciably, they are stimulated to drop back to S_0 as their energy of excitation is used up to manufacture laser photons. The working principle of a dispersive cavity has been described in Chap. 11 in the context of tuning a CO_2 laser, and the same mechanism can also be readily employed here to impart tunability to the emission of a dye laser.

The proximity of the triplet state T_1 to S_1, however, poses a major challenge here and greatly lessens the effectiveness of a dye laser. It is also possible for an excited molecule in state S_1 to cross over to triplet state T_1 before it can be stimulated to drop down to S_0. Every time such an interstate crossing occurs, the pump energy expended to produce the excitation is lost and is not realized as a laser photon. And the story does not end here: The crossover triggers two more ramifications: (1) T_1 is a metastable state, so the population builds up here at the expense of molecules that are available for continuing the lasing process. (2) In many dye molecules, the energy of the laser photons may have a match with the spacing between the second triplet state T_2 and T_1. Hence, the molecules at this metastable state can absorb laser photons and be excited to T_2, and the same is indicated in this figure as triplet absorption. Thus, populating the triplet state is equivalent to creating a laser photon sink within the cavity detrimentally affecting the performance of the laser. There are substances that, when added to the dye, are capable of removing the energy from the T_1 level and releasing the molecules back to the ground singlet state S_0 to further the lasing process. Although system crossing cannot be prevented, its effect will be far less damaging in the presence of such a quenching agent.

10.3.1.1   Dye Laser Pumping

It being a liquid laser, the active length can be much smaller in comparison with that of gas lasers. This makes pumping of the liquid medium possible from the side where the pumping light travels orthogonal to the lasing cavity. Pumping is essentially effected by light of wavelength shorter than the operating wavelength of the laser and can be achieved either by a flashlamp or by another laser.

As the flashlamp produces visible light over a wide range of wavelengths, it is suitable for pumping multiple dyes. For better utilization of the pump photons, the flashlamp and the cell containing the dye solution is often enclosed inside a cylindrical reflector with an elliptical cross section. The pumping arrangement, which is similar to the case of a lamp pumped solid-state laser, has been elaborated later in Sect. 10.4.1. As for laser pumping, the laser to be used is usually dye-specific. Nitrogen, copper vapor, argon ion, and excimer lasers, as well as harmonics of an Nd:YAG laser, often find use as a pump source for dye lasers. If a dye has to generate, say, wavelength tunable red light, it has to be essentially pumped by a green laser. A dye can typically provide tunability of wavelength of several tens of nanometers. Rhodamine 6G, offering approximately 80 nm tunability, appears to be the best. Tuning across a much wider range of wavelength thus calls for switching dyes and consequently pumping lasers as well.

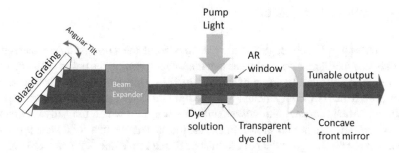

Fig. 10.5   Schematic illustration of the optical layout of the tunable operation of a typical dye laser

10.3.1.2   Working of a Dye Laser

The dye, essentially a long organic molecule, can be dissolved in a variety of solvents, such as ethanol, methanol, toluene, benzene, acetone, etc. or even water, and the solution is enclosed in a watertight cell for its use as a liquid active medium. For low power operation, it is not mandatory to flow the dye solution for heat dissipation. For operations at higher powers, the dye solution is required to be flown to remove waste heat and to prevent rapid dye degradation. As a dye laser is often pumped from the side by either a flashlamp or another laser, the dye cell wall is transparent to allow the passage of the pump light. The optical layout of a typical dye laser cavity with the provision of tuning its wavelength is schematically shown in Fig. 10.5. As seen, the pump light is transverse to the laser cavity. In the case of pumping with a laser, it is quite straightforward to guide the low divergent pump beam through the transparent dye cell. The pumping with a flashlamp, however, is more involved, and the cylindrical reflective housing, prerequisite here to ensure better photon utilization, is not shown in this figure.[7] The cavity is formed with a concave output mirror and a diffraction grating in the Littrow configuration.[8] A beam expander is deliberately introduced between the dye cell and the grating to increase the area of the grating surface illuminated by the intracavity optical beam. This as shown later in Chap. 11 (Eq. 11.21) enhances the resolving power of the grating making it more effective as a tuning device.

---

[7]A qualitatively similar arrangement is also used for the lamp pumping of a solid-state laser and is shown in Figs. 10.7 and 10.8 in a latter section of this chapter.

[8]The working of a diffraction grating, both in the Littrow and non-Littrow modes, has been described in Chap. 11, Sect. 11.3.6.3.2, in the context of tuning the emission of a CO_2 laser.

10.4   Solid-State Lasers

It is now common knowledge that the race to become the very first laser was won by a solid-state medium, a dark horse as a matter of fact, getting the better of the gaseous medium, the favorite. As a rule of thumb, the solid-state medium that performs as a laser usually has two components, viz., the light producing atoms and a passive host medium,[9] often a crystal and occasionally a glass, into which the active species is embedded. There exists no well-defined rule for the naming of a solid-state laser, and this often results in a great deal of confusion. For example, the Nd:YAG laser, one of the most common and widely researched solid-state lasers, is often referred to as the YAG laser, while the same YAG can be the host of many more lasers such as Yb: YAG, Er:YAG, Tm:YAG, etc. that have their own characteristic wavelengths. Or, for that matter, the very first laser, viz., the ruby laser that is essentially a $Cr:Al_2O_3$ laser, was named after the natural gem even though the host here is a synthetic sapphire! Grouping a laser into the solid- state category is also no less ambiguous either; a case in point here is the semiconductor laser. A semiconductor, in the language of physics, is an electronic solid-state device that can conduct electric current; hence, a semiconductor laser should be classified as a solid-state laser. However, in the context of laser physics, the active medium of a solid-state laser is essentially a dielectric and hence does not conduct electricity, so identifying the semiconductor laser as a solid-state laser would be a misnomer. As the active medium is an insulator, a solid-state laser is to be necessarily optically pumped, and it is well known that the ruby laser that Maiman operated in 1960 was pumped by a flashlamp. A semiconductor laser in the jargon of lasers can, if at all, be called an electrically pumped solid-state laser! An elementary knowledge of semiconductor physics is obligatory to understand the mechanism of creating population inversion in a semiconductor when it is driven by an electrical current. It will be prudent to study semiconductor lasers independently and not clubbing with solid-state lasers, and consequently, the working of semiconductor lasers is described in a separate chapter (Chap. 12).

   The discovery of lasing in Cr ions doped in a sapphire crystal led to a proliferation of activity to bring to light more such atom-host combinations for achieving amplification of light. This basically stemmed from the fact that a solid, with inbuilt ability to provide higher density of active species, has the potential to make more powerful lasers. The capability of delivering high power has made solid-state lasers indispensable, in particular, for certain military, industrial, and medical applications.

---

[9]Unlike the gas and liquid lasers, a solid-state laser needs a host that would hold the light emitting active atoms firmly in place and would, at the same time, allow unrestricted passage of both the pump and emitted photons. It is well known that the fraction of the pump energy that is not converted into laser light is realized as heat and the host must have adequate thermal property to allow rapid conduction of this heat to its surface.

Table 10.1 A list of some common solid-state lasers, both tunable and non-tunable

I: Discrete wavelength				
Active species	Host name	Host composition	Central wavelength	Pumping scheme
Chromium	Sapphire (ruby)	Al_2O_3	694.3 nm	Xe flashlamp
Neodymium	YAG	$Y_3Al_5O_{12}$	1064 nm	Flashlamp/diode
	YV	YVO_4	1064 nm	Flashlamp/diode
	GdV	$GdVO_4$	1064, 1341 nm	Flashlamp/diode
	YLF	$YLiF_4$	1043, 1057 nm	Flashlamp/diode
	Glass	Silicate glass	1061 nm	Flashlamp
		Phosphate glass	1054 nm	
Erbium	YAG	$Y_3Al_5O_{12}$	2940 nm	Diode
	YLF	$YLiF_4$	1730 nm	Diode
Holmium	YAG	$Y_3Al_5O_{12}$	2100 nm	Diode
Thulium	YAG	$Y_3Al_5O_{12}$	~2000 nm	Diode
II: Tunable wavelength				
Active species	Host name	Host composition	Spectral range	Pumping scheme
Titanium	Sapphire	Al_2O_3	660–1100 nm	Ar ion, CVL, SH of Nd: YAG
Chromium	Alexandrite	$BeAl_2O_4$	700–825 nm	Flashlamp/diode
Chromium	Emerald	$Be_3Al_2Si_6O_{18}$	730–840 nm	Flashlamp

This table does not include the fiber glass hosts. The fiber lasers have been described in a latter section

It is no wonder that today there exist a wide variety of solid-state lasers yielding coherent light spanning a remarkably broad area of the spectrum. As we shall see below, these lasers can be grossly divided into two main categories, viz., one that basically provides output at discrete wavelengths and the other delivering tunable output. Table 10.1 provides a list of some of the popular discrete and tunable solid-state lasers.

The methodology of operation of most of these lasers is qualitatively similar. To keep the content short and snappy, we would therefore describe here the working of only the Nd:YAG laser that has been the subject of a great deal of research over the years and has emerged as the most popular of all solid-state lasers. The ruby laser has the distinction of being the very first laser but suffers from the disadvantage of being a three-level laser. The neodymium laser on the other hand is a four-level system and thus outperforms ruby and any other three-level lasers. As a matter of fact, so much energy is required to invert the population in a ruby laser that it is not conducive for the *cw* operation. A simplified energy level diagram to describe the working of a Nd: YAG laser is shown in Fig. 10.6; the energy level diagram of a ruby laser is also shown alongside for the sake of comparison. Nd atoms have two dominant absorption bands centered at wavelengths of approximately 700 and 800 nm. The laser is optically pumped, and the active species absorb the incident energy directly without

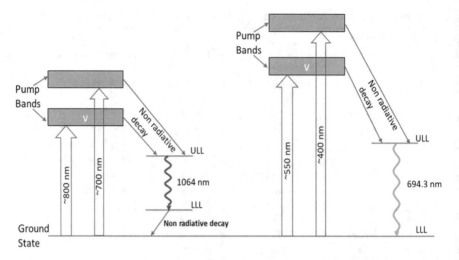

Fig. 10.6 The energy levels that participate in the lasing of Nd:YAG laser is shown in the left, while that for a Cr:sapphire (ruby) is shown on the right. ULL and LLL stand for upper and lower laser levels, respectively

the requirement of an intermediate species as is the case with most gas lasers. The excited atoms then drop down to the metastable upper laser level by rapidly releasing nonradiatively a part of the excitation energy to the crystalline lattice. The lasing transition occurs at 1064 nm in the case of the YAG host. The population arriving at the lower laser level as a result of lasing quickly relaxes nonradiatively to the ground state and becomes available for continuation of the process of lasing.

Historically, the pumping of a solid-state laser by a flashlamp has preceded diode laser pumping by several decades. From the consideration of the working of the laser, irrespective of whether it is lamp or diode pumped, it would always be advantageous to ensure the most efficient utilization of the pump photons. The host material is usually rod-shaped often resembling a small pencil in size. Pumping with a diode has a decided advantage, as it gives out a well-directed beam of light having an adequate spectral match with the absorption of the active species. The broadband emission of a flashlamp, on the other hand, is nondirectional and has poor spectral match with the active medium's absorption. Consequently, as described below, the pump geometry employed in the operation of a lamp pumped solid-state laser differs widely from that for a diode pumped system.

10.4.1 Lamp Pumping

The lamps used for pumping Nd- lasers are usually tubular in shape and run, as shown in Fig. 10.7, along the length of the laser rod. As described earlier in Chap. 7, placing the assembly of the lamp and rod inside a cylindrical reflective enclosure will

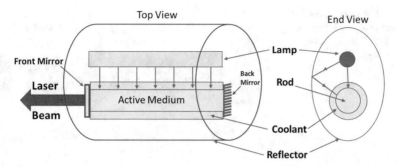

Fig. 10.7 The end and top view of the operation of a typical solid-state laser pumped with a lone tubular-shaped lamp placed parallel to the active medium

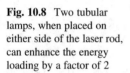

Fig. 10.8 Two tubular lamps, when placed on either side of the laser rod, can enhance the energy loading by a factor of 2

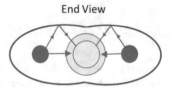

help direct the escaping photons back into the rod to enhance the absorption. Normally, the reflector has an elliptical cross section, and the lamp and the rods are placed along the two foci of the ellipse. This allows a majority of the light emitted by the lamp to be directed by the reflector onto the rod. The usage of two flashlamps to increase the coupling of the pump photons into the active medium is also not uncommon. The cross-sectional (or end) view of such a two-lamp pumping config-uration is illustrated in Fig. 10.8. The reflector is fabricated by fusing two cylindrical enclosures, each with an elliptical cross section. The two ellipses are so joined as to have a common focus along which the rod is placed. As seen in this figure, the two lamps are now placed along the remaining two foci of these two ellipses. Such an arrangement will readily offer an energy loading that is twice as large as that which is possible with a single lamp. A substantial fraction of the pump energy absorbed by the rod will be realized as heat in its interior following the process of lasing. The waste heat that the host eventually conducts away onto its surface is removed by flowing a coolant fluid over it. As in the case of a gas laser, here too the hot fluid discards the waste heat in a heat exchanger. Helical flashlamps, wherein the rod lies along the center of the helix, which powered the very first laser, have not gained much popularity due to hindrances of the positioning of the mechanism for flowing coolant over the hot surface of the rod. Unoptimized cooling of the laser rod leaves its center hotter than the surface which results in the occurrence of thermal lensing

Fig. 10.9 Optical layout of the end pumping scheme of a solid-state laser with a single diode laser

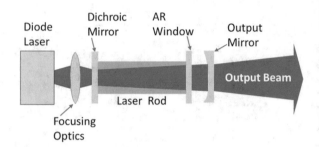

effect[10] and also development of a mechanical stress.[11] While thermal lensing can greatly degrade the spatial quality of the laser beam, mechanical stress may even lead to cracking of the laser rod.

10.4.2   Diode Pumping

As lamp pump solid-state lasers suffer from poor e-o efficiency (~0.5–1%), the wasteful expenditure of energy becomes truly prodigious for a high-power lamp pumped solid-state laser. In contrast when this laser is pumped by the monochromatic and directional emission of a diode laser, it operates with a greatly improved e-o efficiency (10–20%). Not surprisingly, with the advent of efficient diode lasers, the usage of once popular lamp pumping is on the wane, in particular, in the operation of high-power solid-state lasers. In contrast to lamp pumping, the directional emission of a diode laser has also made end pumping of a solid-state laser feasible. As we shall see below, side pumping, however, is inevitable in the operation of lasers with higher power.

10.4.2.1   End Pumping

This simple scheme, also known as longitudinal pumping, where the pump beam is coupled to the laser rod through one of its ends, is used widely for low power lasers. The pumping and cavity configurations are shown in Fig. 10.9. The emission of a single diode laser is optically coupled into the rod through a dichroic mirror that offers high transmission at the pump wavelength and high reflection at the lasing wavelength. The pumping geometry intrinsically offers here a good spatial overlap between the pump photons and the light emitting atoms dispersed uniformly within

---

[10]Thermal lensing effect occurs because the refractive index of the host material depends on the temperature. The hotter center of the rod has higher r.i. compared to that at the edges. A medium with such a spatial inhomogeneity in r.i will focus a beam of light just like a normal convex lens.

[11]Mechanical stress develops as expansion of the hotter center of the rod is higher than at the edges.

the host. This, in turn, yields a high gain making this scheme attractive for use as an optical amplifier as well. The laser rod is usually enclosed between the dichroic mirror and an AR window. In the case of a compact and small laser system with modest power, the output mirror can be directly mounted on the rod in place of the AR window.

The most common example of an end pumped solid-state laser is a green laser pointer that emits at a wavelength of 532 nm. The 808 nm IR emission of a GaAlAs diode laser usually pumps a Nd doped YAG crystal that emits deeper in the IR at 1064 nm. This light is then frequency doubled in a nonlinear crystal producing green light at 532 nm. The overall efficiency of IR to green conversion is approximately 20%; a diode of about 1 W power can thus generate ~200 mW of green light.

As the pump can be introduced here from at most two directions (basically the two ends of the rod), the end pumping schemes are suitable for modest to moderate power solid-state lasers up to several watts or so. The pump power required for the operation of lasers with higher power cannot be provided by one or two laser diodes, and consequently moderate to high-power solid-state lasers are required to be pumped by multiple diode emitters. These lasers are essentially pumped through the sides, and depending on the requirement of power, they can be pumped by a diode bar or a diode array.

10.4.2.2   Side Pumping

The side pumping scheme, also called transverse pumping, of a diode pumped solid-state laser is qualitatively illustrated in Fig. 10.10. The emission from the pump diodes is injected by coupling optics, not shown in the figure, to the laser rod here from the sides. The host crystal, as seen, is surrounded by coolant that flows through an outer glass tube, transparent to the pump light, and the outer surface of the rod. With the arrangement of side pumping, it is readily possible to pump a laser rod with several diode bars each with an emitting region matching closely to the length of the rod. The major advantage of side pumping is that gain, unlike the case with end pumping, remains homogeneous across the length of the rod. The lasers capable of delivering tens of watts to several hundreds of watts can be pumped by diode bars, while the kW class lasers require diode arrays as the pump source.

10.4.3   Tunable Solid-State Lasers

In some solid-state lasers, a very strong interaction between the doped atoms and the host crystal results in an intense coupling of the electronic energy levels with lattice vibrations. Consequently, the optical transition, in addition to the emission of a

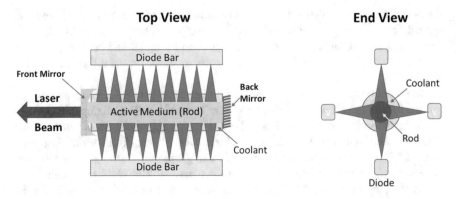

Fig. 10.10 A schematic illustration of the side pumping scheme of a diode pumped solid-state laser. Although two diode bars are shown here for clarity, bars can be used from all sides of the laser rod for better energy loading. The end or cross-sectional view shows pumping from all the four sides

photon, also involves emission or absorption of one or multiple acoustic phonons.[12] This phonon-electronic interaction results in a strong homogeneous broadening of both the originating and terminating electronic energy levels. This thus gives rise to the prospect of tuning the wavelength of these lasers across a range determined by the width of such broadening of the energy levels. As tunability arises from the interplay between lattice vibrations and electronic transitions, these lasers are also called vibronic lasers. The mechanism of the tunability is fundamentally similar to that of the electronic transitions of a molecular gas laser that originate from optical phonon[13]-electronic interactions. As we know, the energy of a vibrating molecule is quantized, and a quantum of energy is equivalent to an optical phonon.

The energy levels that participate in the process of lasing of a vibronic laser can thus be qualitatively represented as shown in Fig. 10.11. The incident pump light excites the atoms from the ground state to the pump band from where they drop down to the upper laser level that basically is a vibronic band. Once in this upper vibronic band, they relax very fast to the lowest energy point of this band. The lasing transition originates from the lowest energy state of the upper band and terminates at any energy location within the lower-level vibronic band. However, the lowest energy state of this band by virtue of being nearest to the ground state will have the maximum population, and the population in this band will be distributed following a Boltzmann distribution. The highest energy state in this band will understandably have the least population. Consequently, the laser gain will decrease with decreasing wavelength. Thus, in a nondispersive optical cavity, lasing will

---

[12] An acoustic phonon is a quantum of atomic lattice vibration that essentially represents a quantum of sound.

[13] An optical phonon is nothing but a photon that represents a quantum of light.

Fig. 10.11 A qualitative energy level diagram of a typical vibronic solid-state laser

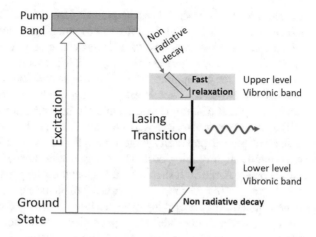

Fig. 10.12 A qualitative representation of the optical layout of the working of a Ti-sapphire laser that is pumped by the frequency doubled output of a Nd-YAG laser

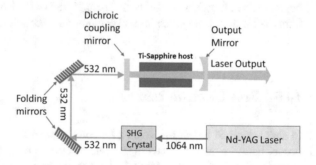

predominantly occur at longer wavelengths. In a dispersive cavity, on the other hand, the wavelength can be tuned across a wide range. The optical arrangement to realize the tuning of the wavelength is qualitatively similar to that used for CO_2 and dye lasers and is described later in Sects. 11.3.6.3.1 through 11.3.6.3.4.

10.4.3.1 Titanium Sapphire Lasers

Among the vibronic lasers in operation today, the Ti-sapphire laser offers the widest tunability covering an incredible range of red (~650 nm) to well into infrared (~1180 nm) of the em spectrum. The extremely strong interaction that titanium exhibits with the host sapphire crystal has made this possible. The Ti-sapphire laser was first operated in 1986, and its growing popularity soon pushed the dye lasers, once ubiquitous in applications requiring a tunable laser, into oblivion. It is appropriate to present here a brief description of the operation of a Ti-sapphire laser, and the same has been schematically illustrated in Fig. 10.12. The construction and operation of a Ti-sapphire laser is quite similar to other kinds of solid-state lasers except that it is challenging to realize diode pumping of this laser. The absorption of Ti-sapphire peaks at around 500 nm at which an appropriate diode laser is not readily available. The second harmonic of the Nd-YAG laser at 532 nm is widely used as a pump source for Ti-sapphire lasers. The frequency doubling of the 1064 nm emission of the Nd-YAG laser is effected in a nonlinear crystal, and the generated coherent light at 532 nm is steered into the host crystal through a dichroic mirror. Although the argon ion laser at 514 nm also finds application as a pump source here, it has not gained popularity largely because of its poor operating efficiency and exorbitantly high running cost. The cavity can be formed here between this and a concave output mirror. It should be noted that the arrangement of wavelength tuning, an integral part of a Ti-sapphire laser, is not explicit in this figure, which is simplified to help understand the working of such a laser. The wide tunability of the Ti-sapphire laser can be readily translated through nonlinear generation of higher harmonics into shorter wavelengths covering a broad range of spectra from blue to UV and even beyond. The ultrawide tunability of Ti-sapphire lasers also allows the generation of ultrashort pulses from these lasers by employing techniques described in Chaps. 8 and 9.

10.5   Free Electron Lasers

In sharp contrast to the operation of a conventional laser, where optical transition occurs between energy levels of discrete electronic bound states of an atom/molecule/ion, in a free electron laser (FEL), the electrons are not bound. The electrons moving close to the speed of light here essentially make transitions between continuous states rather than between two bound states when subjected to a transverse acceleration. The extremely wide tunability of an FEL, spanning from terahertz through infrared, visible, and ultraviolet to X-ray, basically stems from this fact. The operation of an FEL was first conceptualized by John M. J. Madey (1943–2016), an American physicist, in 1970 when he was a doctoral student at Stanford University. Madey is also credited with the development of FEL when in 1975 he and his coworkers at Stanford University succeeded in amplifying the 10.6 μm output of a CO_2 laser by letting it pass through an e-beam coupled 5-m-long undulator. Subsequently, in 1976, they integrated this gain medium with an optical cavity and demonstrated lasing at 3.4 μm wavelength. To gain physical insight into the working of an FEL, it is imperative to take a closer look at the emission of radiation originating from the accelerated relativistic motion of a charged particle.

Fig. 10.13 A charged
particle moving at a
relativistic speed, when
subjected to a transverse
oscillation, emits e-m
radiation within a small cone
in the direction of its motion

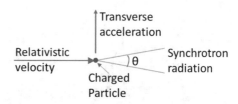

10.5.1 Working Principle

A direct consequence of the Lorentz force law,[14] which together with Maxwell's
equations[15] forms the foundation of classical electrodynamics, is that an accelerated
charged particle will always emit electromagnetic radiation. In the special case when
the charged particle moving at relativistic speed[16] is subjected to a transverse force, it
emits radiation, known as synchrotron radiation, in a narrow cone in the forward
direction. The situation is schematically represented in Fig. 10.13. The angle of the
cone θ is related to the Lorentz factor[17] γ through Eqn. 10.5, where v and c are the
speed of the charged particle and velocity of light, respectively. A popular method to
implement this condition in practice is to propagate an e-beam through a periodic
arrangement of magnets, the poles of which are so arranged as to create a side-to-side
magnetic field. As the Lorentz force causes the electrons to wiggle transversely
following a sinusoidal path (Fig. 10.14), this array of magnets is called an undulator
or a wiggler. The period of the magnetic field λ_0 that essentially is also the
wavelength of the oscillation of the electrons will experience a Lorentz contraction
to appear in the frame of reference of the electron as λ' given by Eqn. 10.6

$$\theta = \frac{2}{\gamma}, \text{where } \gamma = \frac{1}{\sqrt{1 - \frac{v^2}{c^2}}} \tag{10.5}$$

---

[14]Lorenz force, also called electromagnetic force, is the force experienced by a moving charge
particle due to the presence of electric and magnetic fields.

[15]In 1865, James Clerk Maxwell (1831–1879), a Scottish physicist, combined the Gauss's laws of
electricity and magnetism, Faraday's law, and Ampere's law into four equations, aptly termed as a
unified theory of electromagnetic phenomena. The most remarkable inference of Maxwell's
equations, recognized as among the most important in science, is that light is an electromagnetic
wave which manifests as time varying magnetic and electric field propagating in space.

[16]Relativistic speed is the speed at which the relativistic effects become significant. Normally
relativistic effects are accounted for when the speed of an object exceeds one-tenth of the speed of
light.

[17]Lorentz factor is the measure of the change of the physical properties, such as length, and mass, of
a moving body compared to the values at its stationary state.

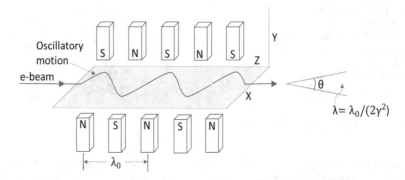

Fig. 10.14 When a relativistic charged particle like an electron is made to travel through an array of alternating magnetic field structure, it forces the electron to periodically oscillate and radiate. The Lorentz contraction and Doppler shift at the relativistic speed cause the initial low frequency of the electron in the rest frame to appear to be extremely high in the laboratory frame

$$\lambda' = \frac{\lambda_0}{\gamma} \tag{10.6}$$

The frequency of oscillation of the electron ν' in this frame is, therefore,

$$\nu' = \frac{v}{\lambda'} \tag{10.7}$$

Upon substituting λ' from Eq. 10.6, we obtain

$$\nu' = \frac{v\gamma}{\lambda_0} \tag{10.8}$$

The electrons oscillating at ν' will also radiate at this frequency, which will be relativistic Doppler shifted[18] in the laboratory frame to

$$\nu'' = \nu' \sqrt{\frac{1 + v/c}{1 - v/c}} \tag{10.9}$$

Combining Eqs. 10.8 and 10.9 and substituting $\gamma = \frac{1}{\sqrt{1 - \frac{v^2}{c^2}}}$., we obtain

---

[18] Relativistic Doppler effect, like its classical counterpart described in Chap. 8, is also a change in frequency of light caused by the relative motion between the source and observer but by taking into account the effects caused by the spatial theory of relativity.

$$v'' = \frac{v}{\lambda_0(1 - v/c)} \tag{10.10}$$

10.5.2 Tunability of Emission

Substituting c/λ for v'', where λ is the wavelength of the emitted radiation in the laboratory frame, we obtain the relation between λ and λ_0 as

$$\lambda = \frac{\lambda_0}{2\gamma^2} \tag{10.11}$$

In the relativistic limit, γ is very high, and hence $\lambda \gg \lambda_0$. This is equivalent to saying that in the relativistic regime, the low frequency of radiation emitted in the electron rest frame is shifted to a very high frequency in the laboratory frame. This equation clearly suggests that the frequency of the emitted light can be readily tuned by varying λ_0, i.e., the undulator period, or γ, i.e., the KE of the e-beam. For instance, for an e-beam energy of 50 Mev, when $\gamma \approx 100$, the emitted light can be tuned from visible to IR by varying the undulator period between 1 and 10 cm. Clearly, any change in the e-beam energy will correspondingly change this tuning window. It should be noted here that the tunability also depends, albeit weakly, on the strength of the magnetic field of the undulator. Furthermore, as the electrons oscillate in the XZ plane, the emitted light will be plane polarized in this plane.

10.5.3 Spectral Broadening of the Emission

As is evident from Fig. 10.14, an observer will see the radiation emitted by the electrons from the time they enter the undulator until their exit from it. If L is the length of the undulator made up of N magnet periods, then

$$L = N\lambda_0 \tag{10.12}$$

The radiation emitted by the e-beam upon entering the undulator will reach an observer standing right at its end after a time $t_1 = \frac{L}{c}$. The observer will continue to see the radiation until the e-beam exits the undulator after a time $t_2 = \frac{L}{v}$. Thus, the light emitted by the electrons will be contained in a pulse of duration

$$\Delta t = t_2 - t_1 = \frac{L}{v}(1 - v/c) \tag{10.13}$$

Combining Eqs. 10.11, 10.12, and 10.13, we obtain

$$\Delta t = \frac{N\lambda}{c} \times \frac{1 - v/c}{1 + v/c} \approx \frac{N\lambda}{c} \text{ (in the relativistic limit)} \tag{10.14}$$

Bandwidth $\Delta\lambda$ of a transform limited pulse is related to its duration Δt by[19]

$$\Delta v \times \Delta t \sim 0.5, \tag{10.15}$$

$$\text{Now, } v = \frac{c}{\lambda}. \text{ Therefore,}$$

$$\Delta v = \frac{c}{\lambda^2} \Delta\lambda \tag{10.16}$$

Combining Eqs. 10.14, 10.15, and 10.16, we obtain

$$\frac{\Delta\lambda}{\lambda} = \frac{1}{2N} \tag{10.17}$$

Thus, the higher the number of magnet periods of the undulator, the more monochromatic the emitted light. Since this broadening mechanism is identical for all photons, the broadening here is essentially a homogeneous broadening. There are a number of factors such as the inhomogeneity in the magnetic field and the angular spread in the direction of motion of individual electrons constituting the e-beam that contribute to the inhomogeneous component of the broadening.

10.5.4   Construction and Operation

We now know that by constructing an undulator of a different magnet period λ_0, it is possible to discretely change the wavelength of the light emitted by electrons oscillating in the periodically varying magnetic field. Furthermore, upon varying either the energy of the e-beam or the strength of the magnetic field of the undulator, the wavelength of this light can be fine-tuned on either side of its emission center. Such a device thus provides the unique prospect of generating coherent light tunable from very long (THz) to very short (VUV and down to even X-ray) wavelengths from a single lasing medium (Fig. 10.15). This distinctive advantage has provided a major impetus to develop a free electron laser by way of integrating this device with a cavity. It is not surprising that the backbone of an FEL is essentially an e-beam undergoing sinusoidal motion in an undulator made up of an array of magnets with their poles alternating periodically. The operation of a free electron laser is schematically illustrated in Fig. 10.16. An e-beam after its production and energy boosting to the relativistic limit in an electron oscillator is injected by bending it

---

[19]Directly follows from Heisenberg's uncertainty relation

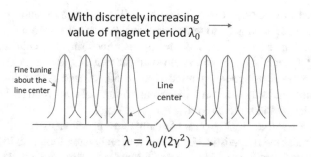

With discretely increasing
value of magnet period λ_0

Fig. 10.15 The undulator can be designed to yield magnet period λ_0 of gradually increasing value that, in turn, will allow generation of light of wavelength λ of discretely increasing magnitude spanning across a very wide region of the e-m spectrum. Varying the e-beam energy or the magnetic field will allow the light to be fine-tuned about the emission line center

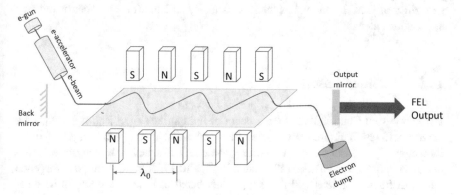

Fig. 10.16 Schematic of a typical free electron laser setup

with a simple dipole magnet into the undulator that makes use of either permanent or electromagnets. The undulator is placed inside an optical cavity. The cavity mirrors are compatible with the wavelength of light that the undulator is designed to emit. The light that the cavity allows to move back and forth inside it stimulates the electrons to emit in phase, causing an intracavity built up of coherent light. A fraction of this emerges through the output mirror as the laser beam. It goes without saying that in the absence of a cavity, the electrons emit in a random manner. A bending magnet is again made use of at the exit end of the undulator to avoid any collision of the electrons with the mirror located here. Instead of dumping the electrons after their passage through the undulator, they are often recycled to boost the working efficiency of the laser. As the electrons move down the wiggler, they lose energy to the growing intracavity light and consequently fall out of resonance condition described by Eq. 10.11. This can be prevented by progressively tapering the strength of the magnetic field of the undulator along its length. This allows satisfactory performance of the laser even when the undulator is much too long.

In addition to its extensive tunability, an FEL also has an inbuilt power scaling capability as there is no need to remove residual or waste heat from the system. An electron gives up a part of its KE toward manufacturing the coherent light and emerges automatically from the system with the rest of the energy retained by it; thus, there is no requirement of removal of residual heat. As we have seen, the removal of heat that stays back into the lasing medium of a conventional laser always poses a challenge and more so for high-power lasers.

Despite the exorbitantly high cost of manufacturing, which essentially stems from the production of the relativistic e-beam and relatively large footprint, there has been a worldwide proliferation of FEL-based programs primarily owing to the unmatched advantages of tunability and power scalability it offers. In addition to its huge potential in basic research, the utility of the FEL has been established in biomedical and photochemical applications. Some of its industrial applications include material processing and laser isotope separation. The inbuilt extreme high-power operation capability of these lasers makes them particularly suitable for a number of military applications as well.

10.6   Excimer Lasers

We know from Chap. 3 that when a beam of light strikes an object, it undergoes a natural diffraction limited divergence that reduces with the wavelength of light. Consequently, a laser beam of shorter wavelength can be focused by a lens to a narrower spot size as has been illustrated in Fig. 10.17. If there were no divergence due to diffraction, a lens will focus a parallel beam of light to a point on its focal plane (top trace). In reality, however, the diffraction limited divergence, which is inevitable, will restrict the focal spot to a finite size. The spot size will be larger for red light than for blue light. The tight focusability, in turn, can make the beam so intense that it ablates the surface directly from the vapor phase, a process known as ablation. While the tiny focal spot generates a well-defined microstructured surface,

Fig. 10.17 A lens will focus a parallel beam of light to a point had there been no diffraction (top). As the beam passes through the lens, it undergoes diffraction resulting in a focal spot of finite size that decreases with the wavelength (shown here as red to blue)

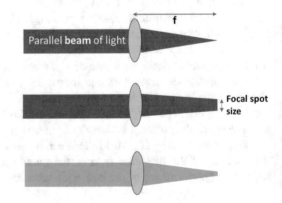

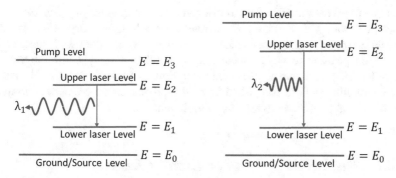

Fig. 10.18 As lowering of the lower laser level is impractical owing to the immovability of the ground level, wavelength of the lasing transition can be reduced only by appropriately raising the upper laser level

material removal through ablation keeps the processed surface extremely clean with no visible thermal effect: one action but two advantages! It is not surprising that this behavior of light makes an ultraviolet laser more suitable for microscale applications such as fabrication of advanced semiconductors, engraving of precision tools, marking of surfaces without changing its structure, or chemical composition and microsurgery. However, with decreasing wavelength, it becomes progressively more difficult to build a laser. This is because as the wavelength of the lasing transition decreases, the upper laser level climbs up in energy, and as a rule of thumb, its spontaneous lifetime therefore continues to decrease. This point can be appreciated by referring to Fig. 10.18, which is basically Fig. 5.9 of Chap. 5 but partially modified in the present context. On the left side of this figure, the lasing transition occurs at wavelength λ_1. In order to reduce the wavelength, one has to look for a lasing species for which the pump and the upper lasing levels climb up as there is no room for the lower laser level to go down. Thus, as the wavelength of the laser decreases, it becomes increasingly difficult to create population inversion, especially for UV lasers characterized by very short wavelengths. To effect a UV transition, the upper laser level is required to be located so high in energy that the population pumped in this level will rapidly leak out through spontaneous emissions. This poses a major challenge to building a laser in the UV or beyond using a conventional gain medium. There are some gas lasers that are capable of emitting UV, but they understandably perform too feebly to serve any meaningful purpose. It is quite appropriate here to consider the case of a vastly popular green laser pointer. Although green is longer than UV in wavelength, it does not originate directly from the lasing medium here. The laser is essentially an infrared laser that is easier to make than a green laser. A fancy crystal is subsequently used to convert the infrared into green light through a nonlinear optical process the mechanism of which will be elaborated in the second volume of this book.

The ability of a rare gas atom to form a molecule only in the excited state rescued the perplexed laser community from the frustration of being unable to build an efficient UV laser. A gainful exploitation of this unique property of the rare gas species to construct a UV laser is credited to Mani Lal Bhaumik (b–1931), an Indian-

American physicist turned philanthropist and a best-selling author.[20] He, during his association with the Northrop Corporate Research Laboratory, Los Angeles, succeeded in creating a Xe_2 molecule in an electron environment that, in turn, served as the gain medium to yield coherent light at 173 nm straight into the vacuum ultraviolet. This work that he presented at the Denver, Colorado, Optical Society of America meeting in 1973 established the credential of such molecules to perform the role of an ideal gain medium for short wavelength lasers.

10.6.1   The Basic Physics of an Excimer Laser

An inert gas atom cannot form a molecule because its outermost shell is completely filled with electrons. When two such atoms approach each other, a repulsive force sets in that, in turn, pushes and keeps them apart inhibiting the formation of a molecule. However, there will be a complete turnaround of the situation if these atoms are first electronically excited and then made to interact. As the outer shell is now partially filled, they will now pull each other to form a stable molecule in the same way that it happens during the formation of a conventional molecule. The resulting molecule is essentially an excited dimer and therefore abbreviated as "excimer." Let us be more specific here and consider Xe as our rare gas atom. The excimer of Xe, which can be denoted as $Xe_2{}^*$, is none other than an electronic excited state of the molecule Xe_2, the ground electronic state of which is so short-lived (~psec) that it is considered to be nonexistent for all practical purposes. (This essentially means that the creation of even a single $Xe_2{}^*$ molecule will result in a population inverted condition.) However, $Xe_2{}^*$ cannot exist indefinitely with its excitation energy and releases it as a UV photon by making a quantum jump to its ground state that disintegrates into two Xe atoms in a flash. This spontaneously emitted photon can initiate stimulated emissions and, in turn, lasing in these excited Xe dimers as the empty ground state allows realization of population inversion and its sustaining with minimum effort. The situation has been schematically described in the illustration of Fig. 10.19. It needs to be emphasized here that just as an electronically excited Xe combines with another excited Xe to form its excimer, it can also readily form a molecule with an atom of another species that is not inert.

---

[20] Mani Lal Bhaumik was born in a poverty stricken Bengali family and narrowly escaped death in the 1942 Bengal famine when over three million Bengalis perished primarily out of starvation and malnutrition. His journey from a mud hut in an obscure Bengal village to the marble mansions of California is a genuine story of rags to riches, indeed a leaf straight out of a fairy tale. As a child, he walked barefoot every day to his village school and worked hard to top the high school examination and earned a scholarship to attend college in Calcutta where he acquired his first pair of shoes in life. His invention of excimer laser made possible the LASIK eye surgery that is now extremely popular across the globe and laid the road for him to walk to the pinnacle of fame. His book, *Code Name God: The Spiritual Odyssey of a Man of Science*, represents an endeavor toward bridging the divide between science and spirituality. The profound impact that this book has made on humanity is evident from its translation into over 100 languages!

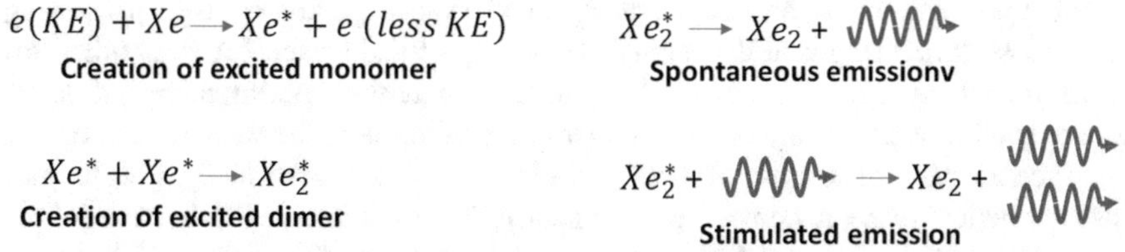

$$e(KE) + Xe \longrightarrow Xe^* + e \; (less \; KE)$$
Creation of excited monomer

$$Xe_2^* \longrightarrow Xe_2 + \text{〰〰}$$
Spontaneous emissionv

$$Xe^* + Xe^* \longrightarrow Xe_2^*$$
Creation of excited dimer

$$Xe_2^* + \text{〰〰} \longrightarrow Xe_2 + \text{〰〰〰〰}$$
Stimulated emission

Fig. 10.19 Electron upon colliding with a Xenon atom loses part of its KE, and the atom, in turn, gets electronically excited. Two excited Xenon atoms can combine to form an excited dimer of Xenon. The excited Xenon molecule, with a radiative half-life of about a nanosecond, spontaneously decays to the ground state by emitting a UV photon that can stimulate another excited Xenon molecule to emit and so on

Table 10.2 Rare gas dimers

Name	λ (nanometer)
Xe_2^*	173
Kr_2^*	146
Ar_2^*	126

Table 10.3 Rare gas halides

Name	λ (nanometer)
XeF	351
XeCl	308
XeBr	282
KrF	248
ArF	193

Calling this molecule an excimer will be a misnomer as it is more complex than a simple excited dimer, and, although less common, lasers based on such variants of the excimer are called exciplex lasers. As a matter of fact, noble gas halide excimer lasers are more popular from an application point of view. They are also high repetition rate compatible and capable of delivering both higher average power and energy per pulse. The wavelengths of operation of some excimer lasers belonging to both rare gas dimers and rare gas halides are depicted in the following tables.

As seen, while the excimers based on dimers usually operate at shorter wavelengths lying in the VUV region of the e-m spectrum, rare gas halide lasers, with the only exception of ArF, operate in the UV region. It is also important to note here that rare gas halide-based excimer lasers are relatively more energetic than their rare gas dimer variants. The popular rare gas dimer and halide lasers along with their wavelengths have been listed in Tables 10.2 & 10.3.

10.6.2   Operation of a Noble Gas Halide Laser

The majority of the excimer lasers sold today are meant for use either in medicine, primarily eye surgery involving the cornea, or in the fabrication of integrated circuits by photolithography in the semiconductor industry. Energetic ArF lasers possessing the shortest wavelength in the family of noble gas halide lasers are best suited for both of these applications. Although we describe here the working of an ArF laser, the operation of a majority of the excimer lasers will follow the same methodology.

The ArF laser makes use of three gases for its operation: Normally a buffer gas, usually helium or neon, constitutes the majority (approximately 90% or more) of the gas mixture, and argon and fluorine together form the rest of the mixture. The buffer gas does not directly participate in the process of lasing but, as we shall find later, plays a crucial role in the operation of the laser. The electron environment that is a prerequisite for manufacturing the electronically excited ArF molecule is usually created by subjecting the gas mixture to a high voltage discharge. The lasing cycle begins when kinetically energetic electrons present in the discharge collide with both ground state Ar atoms and F_2 molecules. The collisions result in both excitation and ionization of Ar atoms and the splitting of F_2 molecules into constituent atoms together with the formation of negative ions as shown below:

$$e + Ar \rightarrow Ar^* + e \qquad (10.18)$$

$$e + Ar \rightarrow Ar^+ + 2e \qquad (10.19)$$

$$e + F_2 \rightarrow F + F^- \qquad (10.20)$$

The presence of buffer species comes handy at this point, as they perform the role of a third body to facilitate further collisions among these freshly generated species, leading to the formation of electronically excited ArF molecules. The sequence of collisions is nevertheless complex, and the two likely channels for the formation of ArF* are

$$Ar^+ + F^- + He \rightarrow ArF^* + He \qquad (10.21)$$

and

$$Ar^* + F_2 \rightarrow ArF^* + F \qquad (10.22)$$

In the former reaction, wherein the argon and fluorine ions combine to form a single species, viz., ArF*, the buffer atom helium performs the role of the third body to allow momentum conservation. The above kinetic processes are more probable at higher gas pressure as it presents a better collisional environment and consequently the excimer lasers operate at high pressures usually exceeding an atmosphere. As explained in Sect. 7.5.2.2 of Chap. 7 (and also elaborated in the following chapter dealing with the working of CO_2 lasers), stabilizing a high-pressure gas discharge is not possible in the *cw* mode. It is, therefore, mandatory to operate an excimer laser in the pulsed mode. As a matter of fact, a discharge that is transverse to the optical cavity, much like in the case of the TEA-CO_2 lasers, is essential to transfer the energy stored in the discharge condenser homogeneously into the interelectrode gaseous volume. The methodology to achieve this is qualitatively similar to that employed in the operation of a high-pressure CO_2 laser and elaborated in Sects. 7.5. 2.2 and 11.3.3. There is, however, a major difference; the discharge duration here is much shorter, usually a few tens of ns, compared to hundreds of ns, which results in the most optimized performance of conventional TEA-CO_2 lasers. For the typical operating conditions of an excimer laser, the discharge tends to degenerate into an arc if its duration exceeds 50 ns or so. Realization of a discharge of such short duration is nevertheless challenging, as laying down the electrical components and designing the laser head must be conducive to an extremely fast buildup of current. The phenomenon of saturation of a magnetic core is often exploited to appropriately compress the current pulse before it is fed to the laser head to ensure an arc-free discharge. Handling and disposal of the extremely corrosive halogen gases, the use of which is mandatory in the operation of inert gas halide lasers, is another major challenge here. The entire laser head, which primarily includes the lasing chamber, the discharge electrodes, and the gas flow recirculatory system comprising a heat exchanger and gas blower, must therefore be made out of halogen compatible materials. Needless to say, building an excimer laser is both more involved and expensive in comparison with most conventional gas lasers.

Striking a rapid discharge lasting just a few tens of ns is undoubtedly an uphill task, but it offers an additional benefit as far as the performance of the laser is concerned. The rapid pumping causes the population inversion to build up at a faster rate allowing stimulated emissions to quickly realize the inversion as the laser output. A slower rate of pumping, on the other hand, would allow the population from the short-lived upper laser level (radiative lifetime ~10 ns) to escape substantially through spontaneous emissions before the stimulated emission rate can appreciably build up to realize them as coherent light. Thus, an excimer laser would most certainly underperform even if striking a glow discharge were possible here with a current pulse of longer duration as in the case of a CO_2 laser.

Excimer lasers intrinsically possess high gain primarily for the following two reasons: A rapid discharge, mandatory for achieving a glow discharge, causes the excited level population to build up rapidly, and an excimer or, more aptly, an exciplex after making a transition from the excited to the ground state dissociates almost instantly. The ensuing high gain allows excimer lasers to lase with no requirement of a high Q cavity, which is normally mandatory for conventional gas lasers. As a matter of fact, an excimer laser is able to perform optimally with a fully reflective mirror at the rear end and an uncoated glass window at the front end that reflects only a small fraction of light back into the cavity. The duration of the laser pulse that is basically governed by the rise time and magnitude of the current pulse as well as the dynamics of molecular kinetics represented by Eqs. 10.18 through 10.22 can vary from several ns to tens of ns. The typical energy per pulse that can vary from a few millijoules to about a joule can show a gradual reduction with increasing pulse repetition rate. Although repetition rates of tens of Hz to about a 100 Hz are more common, the repetitive operation capability of excimer lasers can extend up to several kHz. Although more expensive but better suited for photolithography, the kHz repetition rate will be possible only if the power supply can deliver the required current and the gas recirculatory unit can remove the dissipated heat. The electric discharge pumped excimer lasers can operate with a wall-plug efficiency between 0.2% and 2% and thus perform more efficiently compared to most other gas lasers. The electron beam pumped excimer lasers can yield even higher wall-plug efficiency.

10.7   Chemical Lasers

In early 1983, when Roland Reagan was president, America initiated a program aimed at developing an air defense system to dispel the threat of a nuclear attack by way of rendering the nuclear weapons obsolete. This ambitious plan relied heavily on high-power airborne lasers[21] with the capability of striking a nuclear warhead fired from anywhere in the globe. The idea was basically to destroy the missile in the boost phase, just after the launch, in the sky of the country of its origin. To this end, a chemical laser that does not require electricity for its operation and is capable of delivering a light beam of colossal intensity emerged as the most suitable for deployment in an Earth orbit. Although this scheme, known as the "strategic defense initiative" (SDI), fizzled out in the early 1990s, primarily owing to the end of the Cold War, the utility of chemical lasers as a light-based lethal military weapon is recognized across the globe. As a matter of fact, the major thrust of the development of chemical lasers stems from their potential to defend against both short and long range missiles. To the intense beam of light, fired from the laser, that can traverse the perimeter of the Earth in just a tiny fraction of a second, the missile, however much faster it moves, would appear to be frozen; the end result is thus a precision strike.

A chemical laser is a molecular gas laser that operates in the infrared region and, unlike a conventional laser, is powered by a chemical reaction. The possibility of a chemical reaction pumped laser was conceptualized by a Germany-born Hungarian-Canadian chemist John Charles Polanyi (b–1929) in 1961 soon after the invention of laser in 1960. He realized the prospect of the excess energy liberated in a chemical reaction to vibrationally excite the molecules formed in the process. The pioneering work on chemical reaction dynamics and kinetics eventually fetched Polanyi the 1986 Nobel Prize in Chemistry. American scientists Jerome V. V. Kasper (b–1940) and George C. Pimentel (1922–1989) operated the first chemical laser at the University of California, Berkeley, in 1965 [33]. They were able to initiate a chemical reaction between hydrogen and chlorine with a flashlamp. The reaction yielded hydrogen chloride (HCl) molecules in a vibrationally excited state, automatically establishing population inversion, a condition prerequisite for lasing. Subsequently, the operation of a variety of chemical lasers has been demonstrated, and HF/DF lasers have, in particular, attracted the attention of military developers. The

---

[21] President Reagan, a vocal supporter of the SDI, advocated that "lasers in space" would be a tool to rid the world of nuclear weapons and did not rule out the possibility of eventually sharing it with erstwhile Soviet Union.

operation of the HF laser was first reported in 1967, and rapid development followed thereafter and, by 1982, HF lasers capable of delivering *cw* power exceeding MW were demonstrated. The working of an HF laser has been elaborated in the following section.

10.7.1 Hydrogen Fluoride (HF) Lasers

The HF laser is an infrared laser, and the lasing originates from the transitions between the rotational vibrational levels of the ground electronic state of the molecule akin to the operation of a CO_2 laser. (It is a good idea at this point to study the basics of molecular spectroscopy described at an elementary level in Sect. 11.2 of the following chapter.) Vibrationally excited hydrogen fluoride molecules are produced when atomic fluorine reacts with molecular hydrogen:

$$F + H_2 \rightarrow HF^* + H \tag{10.23}$$

HF^* represents a vibrationally excited state of the product molecule. Fluorine is made available in its atomic form usually by striking a high voltage electric discharge, much the same way in the operation of a CO_2 laser, in the SF_6 gas in the vicinity of the lasing zone. Upon colliding with the kinetically energized electrons present in the discharge, the SF_6 molecules dissociate and liberate atomic fluorine:

$$SF_6 + e(\text{KE}) \rightarrow SF_5 + F + e(\text{less KE}) \tag{10.24}$$

The first four vibrational levels, viz., $v = 0, 1, 2,$ and 3 states of the hydrogen fluoride molecule, are found to be populated as they are formed through fluorine hydrogen reactions. The relative rates of their chemical reaction driven pumping have been measured to be $v_3 : v_2 : v_1 : v_0 :: 12 : 20 : 6 : 1$ yielding, as shown in Fig. 10.20, their respective populations approximately as 31%, 51%, 15%, and 3%

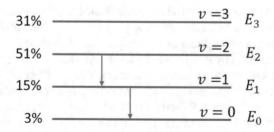

Fig. 10.20 The first four vibrational levels of the HF molecule along with their approximate populations are shown. The tunability of the HF lasers originates from the rotational vibrational transitions between E_2 and E_1 and E_1 and E_0 energy levels

of the total number of *HF* molecules formed. Clearly, population inversion occurs between E_3 & E_1, E_3 & E_0, E_2 & E_1, E_2 & E_0, and E_1 & E_0. However, the transitions between E_3 & E_1, E_3 & E_0, and E_2 & E_0 violate the selection rule, as they result in a change in the vibrational quantum number by more than 1. Lasing in the *HF* chemical laser therefore results from the transitions between E_2 and E_1 and E_1 and E_0. As all the vibrational levels have their rotational manifolds, an *HF* laser, therefore, much like CO_2 lasers, can operate on many discrete wavelengths spanning 2.6–3.0 μm corresponding to different rotational vibrational transitions. Using the stable rare isotope of hydrogen, viz., deuterium, the lasing window shifts to 3.6–4.0 μm. Although the atmosphere is more transparent at these wavelengths, the *DF* laser is still not very attractive for military applications as with increasing wavelength, the focal spot intensity decreases.

10.7.2   *Chemical Oxygen Iodine Lasers (COIL)*

It is worth mentioning in this context of yet another electronic transition chemical laser, popularly known as COIL, that operates at a much shorter wavelength, viz., 1.3 μm wavelength, at which atmosphere exhibits good transmission. This in conjunction with the tighter focusability at shorter wavelength makes this laser more effective in destroying military targets than the longer wavelength *DF* laser. The chemical oxygen iodine laser, abbreviated as COIL, takes advantage of a chemical reaction that produces oxygen molecules in a metastable excited state. These long-lived excited oxygen molecules subsequently dissociate iodine molecules producing electronically excited iodine atoms which when enclosed in a cavity generate a powerful beam of light.

10.8   Gas Dynamic Lasers

In a gas dynamic laser, which can be considered a variant of chemical lasers, the thermal energy, instead of chemical energy, stored in a gaseous mixture is maneuvered to power the laser. The different relaxation rates of the various energy levels of a gas are exploited here to create a population inverted condition and, in turn, lasing by subjecting the gaseous system to rapid expansion through a nozzle. The scheme was first conceptualized by Abe Hertzberg and Ian Hurley, whose attempt in 1965 to establish population inversion between electronic energy states of xenon gas was, however, unsuccessful. The invention of a gas dynamic laser is, however, credited to Edward Gerry and Arthur Robert Kantrowitz (1913–2008) of Avco-Everett Research Laboratory, Massachusetts, USA. They succeeded in inverting the population between energy levels of molecular vibrations and operated the first gas dynamic CO_2 laser in 1966. The operation of a CO gas dynamic laser was subsequently realized in 1970 by Robert L. McKenzie at the Ames Research Center, NASA, California, USA. The capability of a gas dynamic laser to generate quasi *cw*

Fig. 10.21 A 2D schematic view of the working of a typical gas dynamic laser

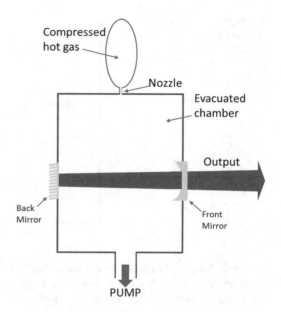

power to the tune of tens of kW, with potential of exceeding even 100 kW, makes them particularly suitable to serve as high-power laser weapons for military applications. With regard to industrial or civil applications, gas dynamically pumped lasers have not generated much interest as they are unable to match the construction and operational simplicities of their electrically pumped counterparts. Here, we describe the operation of a gas dynamic CO_2 laser to gain physical insight into the working of a gas dynamically pumped laser in general.

10.8.1   *CO₂ Gas Dynamic Lasers*

To understand the basics of a gas dynamic laser, we consider a collection of hot CO_2 gas at high pressure. The operation of a typical gas dynamic laser is represented very qualitatively in the illustration of Fig. 10.21, which is easy to understand. At an elevated temperature, a significant fraction of the molecules will move away from the ground state to the higher vibrational states following a Boltzmann distribution. We shall consider only the low-lying vibrational levels from symmetric, bending, and asymmetric modes of vibrations that participate in the process of lasing,[22] and the same is shown in Fig. 10.22. With the gas being at a thermal equilibrium, the population at the upper laser level 001 will be lower than the population residing on any one of the lower laser levels 100 and 020. The hot and compressed gas is then made to expand through a tiny nozzle into an evacuated chamber. As the expanding

---

[22] The working of a CO_2 laser has been described quite extensively in Chap. 11.

Fig. 10.22 The low-lying
vibrational levels of the CO_2
molecules that directly
participate in the process of
lasing

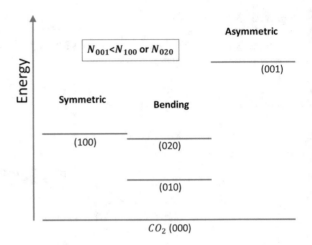

gas cools, the population residing in the energy levels of each mode of vibrations can relax within itself through resonant *V-V* energy transfer collisions. As an example, we consider two CO_2 molecules, one in third and another in second asymmetric vibrational mode, and study their relaxation as they collide with themselves and ground state molecules:

$$CO_2(003) + CO_2(002) = 2CO_2(002) + CO_2(001) = CO_2(002)$$
$$+ 3CO_2(001) = 5CO_2(001) \tag{10.25}$$

The relaxation through *V-V* energy transfer thus eventually culminates in five CO_2 molecules in the first asymmetric vibrational mode. Clearly, all the molecules in this asymmetric stretch ladder, regardless of their level of excitation, will, as the gas cools, relax rapidly to the first asymmetric vibrational mode. Relaxation of the molecule from here to the ground state will involve transformation of vibrational energy to translational energy and is an extremely slow process. The rate of such *V-T* relaxation processes decreases with increasing vibrational energy. Similar resonance *V-V* relaxation will also be operative in the symmetric and bending modes of vibrations as well. However, as there is a close match between the 100 and 020 levels, the molecules in the 100 level can make it to the first bending vibrational mode, viz., 010, through the following collisions:

$$CO_2(100) + CO_2(000) = 2CO_2(010) \tag{10.26}$$

With the level 010 being much closer to the ground state, the *V-T* relaxation from here will be much faster compared to the 001 level. In a nutshell then, as the expanding gas cools, the lower laser levels will depopulate much faster than the upper laser level, quickly establishing a population inverted condition. Construction of an optical cavity transverse to the expanding gas at an appropriate downstream location will make a CO_2 gas dynamical laser operational. The addition of N_2 and *He* remarkably improves the performance of the laser, just as in the case of the conventional operation of a CO_2 laser. N_2, as we know, acts as a reservoir of vibrational energy and resonantly transfers it to the upper laser level by colliding with ground state CO_2 molecules:

$$N_2(V=1) + CO_2(000) = N_2(V=0) + CO_2(001) \qquad (10.27)$$

Helium, on the other hand, plays the role of a quencher as it helps depopulate the lower laser level by efficiently removing the population from the 010 level through V-T relaxation:

$$He(KE) + CO_2(010) = He(\text{More KE}) + CO_2(000) \qquad (10.28)$$

The gaseous mixture pumped out following the process of lasing is available for the next cycle of operation.

10.9   Fiber Lasers

A key performance parameter of the laser is undoubtedly its power generation capability. Understandably therefore, efforts to push up the power output from a laser have been continued ever since its invention in 1960. We know that a fraction of the pump power is always dissipated as heat in the active medium. Removal of this heat is a challenging task as the active medium of a laser is almost always characterized by a rather small surface area to volume ratio and often limits the extractable power from the laser. In a gaseous or liquid laser, thermal management is less difficult as the waste heat can be dissipated by flowing the hot fluid through a heat exchanger. A similar heat removal mechanism is impossible in a solid-state laser, and as we know from Sect. 10.4.1, with increasing power, the laser rod invariably suffers from the effect of thermal lensing that grievously affects the spatial quality of the output beam. The only exception is the fiber laser which makes use of a gain medium similar to that of solid-state lasers but varies widely in geometrical shape. While the active medium of a solid-state laser is usually a light emitting atom doped glass rod typically several centimeter in diameter and tens of centimeter in length, lasing in a fiber laser also originates from a similarly doped glass but in the form of a fine strand, several micrometer to millimeter in diameter and several meter to even up to many kilometer long. The thin and long geometry readily offers a remarkably high surface to volume ratio, and thus, the fiber lasers have inbuilt capability of rapid dissipation of heat to the surroundings. The fiber laser was invented in 1961, within a year of the invention of laser, by American scientist Elias Snitzer (1925–2012) when he optically pumped a strand of glass fiber by wrapping it around a flashlamp. However, extremely inefficient pumping of a fiber through the enormously large later surface area did not allow the exploitation of its heat dissipation capability. This limitation largely halted any meaningful progress in the research and development of fiber lasers for almost three decades. Fiber lasers had their second birth in the 1990s once heterostructured diode laser technology matured enough to allow the end pumping of a fiber. Around this time, British

Fig. 10.23 When light shines on a rarer medium from a denser medium, the rays that are incident at an angle greater than the critical angle θ_c are totally reflected. All other rays are partly reflected and partly refracted

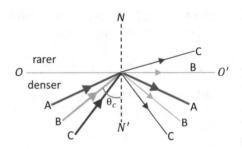

physicist David Neil Payne (b–1944) and his coworkers at Southampton University invented a fiber-based light amplifier. The capability of such an amplifier to directly boost the greatly weakened optical signal that has travelled through a long distance of optical fiber without requiring it to be converted first into electrical signal brought a revolution[23] in fiber-optic communication and Internet connectivity. This has played and is still playing a pivotal role in the research and development of both fiber laser and amplifier systems. Thanks to the pioneering work of British-American scientist Charles Kao,[24] today we have optical fibers through which light, packed with digital data, can travel with practically no loss. Ready scalability of the power of the fiber lasers, which has reached today tens of kW in *cw* and MW in pulsed modes, owes primarily to this. We shall study here at an elementary level the mechanism of pumping and working of a fiber laser. The working of a fiber amplifier that has made possible the operation of an all optical communication system and, in turn, brought a revolution in the fiber-based telephone and Internet connectivity today will be covered in the next volume of this book.

10.9.1   End Pumping of a Fiber Laser and the Confinement of Light

The ability of an optical fiber to keep the light confined within its core and a large acceptance angle for the entry of light through its ends are the key to the emergence of fiber lasers as the leader in industrial, medical, and directed energy applications. An optical fiber acquires these abilities by exploiting the property of the total internal reflection of light that we have studied earlier in Chap. 2. It may be appropriate here to take another look at the phenomenon of total internal reflection of light the occurrence of which has been illustrated as a function of angle of incidence in Fig. 10.23. Light shines here from a denser to rarer medium, separated by the

---

[23] The profound impact of *all* light fiber-optic communication, without the requirement of any intermediate conversion of light into electronic form and back to light again after amplification, on humanity will be elaborated in detail in V-II of this book.

[24] Nicknamed as the "father of fiber-optic communication," Charles Kuen Kao (1933–2018) won the 2009 Nobel Prize in Physics for his pioneering work that laid the foundation for the development of nearly lossless fiber that exists today.

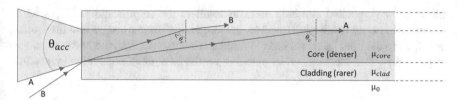

Fig. 10.24 A small section of a typical fiber shown here is not drawn to scale. Usually, the diameter of the core of the fiber (several micrometers) is much too small compared to its length (several meters). As shown, the incident beam, when fits into the acceptance angle θ_{acc}, will be confined into the fiber core and travel down its length without any spillage

interface OO'. Of the three rays shown, ray B is incident at the critical angle θ_c, while rays A and C are incident at angles greater than and lesser than θ_c, respectively. Understandably, therefore, rays A and C are partly refracted and partly reflected, while ray A undergoes total reflection with no refraction. We may therefore conclude that any ray of light that lies between B and the interface OO' does not lose energy through refraction and exhibits complete reflection. Exploitation of this fact by an optical fiber has made possible the operation of a fiber laser. An optical fiber in its simplest form, as shown in Fig. 10.24, consists of an internal glass core and a cladding of a lower refractive index that surrounds the core. The acceptance angle θ_{acc} is the maximum angle that an incident beam of light can be imagined to be subtending at the fiber axis in order to be able to be trapped inside the core. It essentially is a function of the refractive indices of the core (μ_{core}), cladding (μ_{clad}), and the surrounding medium (μ_0) and can be expressed as

$$\theta_{acc} = \sin^{-1}\left(\frac{1}{\mu_0}\sqrt{\mu_{core}^2 - \mu_{clad}^2}\right) \qquad (10.29)$$

As can be seen in this figure, ray A, which marks the lower boundary of the incident light beam, after its refraction into the core strikes the interface of the core and cladding at the critical angle of incidence θ_c. Although not explicitly shown, it is easy to visualize that the ray that defines the upper boundary of this beam will strike the core-clad interface at θ_c as well. The end result is thus the confinement of the entire incident conical beam of light, represented in 2D here, inside the fiber core. To drive this point home, the trajectory of another ray B, not part of this beam of light, as it travels through the fiber is also depicted here. It strikes the core-clad interface at an angle smaller than θ_c and consequently does not undergo total internal reflection, and a part of its energy is spilled out through refraction.

Guided by this judiciously cladded structure, the input pump beam travels down the fiber core and energetically excites the light emitting molecules doped in the pure glassy core of the fiber. A fraction of the photons spontaneously emitted by these excited dopant molecules that strike the core-clad interface at angles exceeding θ_c will be trapped and guided in the core, as elaborated in Fig. 10.23, by total internal reflection. These spontaneous photons understandably perform the role of seed photons for stimulated emission and can result in the amplification of light with appropriate cavity feedback. However, before gaining deeper insight into the working of a typical fiber laser, we need to familiarize ourselves with the basic spectroscopy of the active medium and its pumping for the creation of population inversion.

10.9.2   The Active Species in a Fiber Laser

In the operation of both the fiber laser and fiber amplifier, the optical fiber functions as a passive element that does not produce light. They perform the role of glass host into which the light producing active molecules are embedded akin to the doping of the glass rod in the case of a conventional solid-state laser. However, quite unlike these lasers for which the most common dopant is neodymium, which emits light at 1.06 μm, ytterbium is the more popular lasing molecule in the case of a fiber laser. Ytterbium-doped fiber lasers are the most powerful and emit both *cw* and pulsed light that is tunable over 1.02–1.12 μm. Although the very first fiber laser made use of neodymium molecules as its active medium, this has not gained popularity primarily because of the following reason: By the late 1980s, when there was a revival of interest in the development of fiber lasers, neodymium-based solid-state lasers producing coherent light at 1.06 μm were already in use for many industrial and scientific applications. The other dopant molecules that find applications in the fiber lasers include erbium (1.5 and 2.9 μm), thulium (1.47 and 1.8 μm), and holmium (2.1 and 2.9 μm). While Er-doped lasers lag behind the Yt-doped ones in the power producing capability, Er-doped fiber amplifiers (described in Sect. 10.9.5), on the other hand, have revolutionized the way in which the information is transmitted globally. Thulium- and holmium-doped lasers with wavelengths much longer than 1.4 μm fall into the so-called eye-safe region of optical wavelengths and are of interest owing to their potential in many military and industrial applications.

   All these dopants that perform as the active species for the fiber lasers, however, have qualitatively similar energy level structures. As a general case, we therefore focus our attention to the energy level representation of ytterbium-doped fiber laser. The energy levels that participate in the process of lasing are shown in Fig. 10.25. As seen, both the excited and the ground energy states are split into multiple closely spaced sublevels owing to the interaction between electrons and the electric field present within the solids known as the Stark effect. As the majority of populations reside in the lowest Stark level of the ground state, absorption essentially occurs from here. In addition to the two strong absorption peaks at 915 and 975 nm, there exists a weak absorption at ~860 nm. The population excited to any one of the upper state sublevels relaxes nonradiatively through multiphonon decay eventually to the lower-most Stark level of the excited state which is a metastable state with a lifetime of around 1 ms. The transitions originating between the excited state and all the four stark levels of the ground state manifold result in lasing around 1140, 1090, 1035, and 975 nm. As the lower levels for the 1140, 1090, and 1035 nm transitions are sparsely populated at ambient temperature, the fiber laser acts here like a four-level

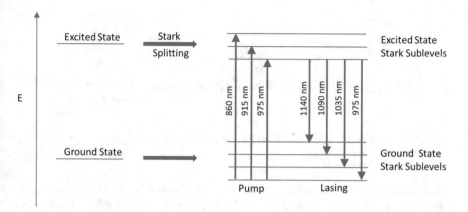

Fig. 10.25 Energy level diagram showing schematically the popular pump and lasing transitions of ytterbium-doped fiber laser

laser. The 975 nm transition, on the other hand, terminates on the vastly populated lowest stark level of the ground state, and the fiber laser here therefore acts like a three-level laser. A major consequence of the Stark splitting of the excited and ground electronic states is thus the occurrence of gain and, in turn, lasing across a wide range of wavelengths.

10.9.3   The Pumping of a Fiber and Its Lasing

Like any other solid-state laser, fiber lasers are also optically pumped. Although the very first fiber laser, operated in 1961, did make use of side pumping, their long strand-like shape makes these lasers unsuitable for pumping through the side. Furthermore, the difficulty of coupling the divergent light emitted by a conventional source through the miniscule end of the fiber was primarily responsible for putting any serious research on fiber lasers on hold during the 1960s and 1970s. The advent of heterojunction diode lasers in the 1980s offered the prospect of transporting light efficiently through the end of a fiber and led to the resurgence of interest in the research on fiber lasers. There has been no stoppage thereafter, and these lasers have now exploded out of the laboratory into numerous applications profoundly impacting humankind.

We know that for end pumping to be effective, the light must enter the core of the fiber within its acceptance angle, a requirement impossible to be met by conventional lamps or LEDs. These sources are usually much too large and emit into an angle that is too wide to allow any meaningful coupling of the light into the fiber core. The diode lasers, on the other hand, emit from a tiny area into a much narrower angle that can be effectively transported into the fiber core through its end. Once inside, the pump light, as we know, will be guided through the length of the core by the

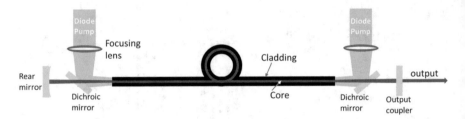

Fig. 10.26 A simple cavity configuration of a fiber laser which is end pumped by the emission of diode lasers from both sides

phenomenon of total internal reflection. For a diode laser of appropriate wavelength, the light will be gradually absorbed by the doped molecules as it travels down the fiber. For example, the two strong absorptions of ytterbium can be accessed by diode lasers operating around 915 and 975 nm. Once the fiber has been made active by inverting population, it can perform as an amplifier. However, like any other laser, the onset of lasing in an active fiber requires a cavity. A very simple fiber laser cavity configuration along with the end pumping arrangement from both sides is schematically illustrated in Fig. 10.26. The emission from the diode laser is appropriately focused and then launched into the fiber by means of suitably located dichroic mirrors that offer high reflection at the pump wavelength and high transmission at the lasing wavelength. The pump and laser beams are indicated in green and red, respectively, to underscore the fact that lasing always occurs at a wavelength longer than the pump. As seen, the optical cavity is formed by a high reflectivity rear mirror and a moderately reflective output coupler. The central circular twist on the fiber seen in this figure basically emphasizes that the active length here usually far exceeds the cavity length. The fiber is therefore coiled so that the whole apparatus fits into a conveniently sized enclosure. The ends of the fiber are antireflection coated to avoid intracavity loss through Fresnel reflection. The optical cavity formed with such external mirrors is, however, quite prone to becoming misaligned due to the occurrence of thermal changes or the presence of physical bumps in the fiber that are inevitable. The alignment problem can be readily overcome by fabricating appropriate Bragg mirrors inside the optical fiber at its two ends, quite akin to the usage of Bragg reflectors in the operation of vertical cavity surface emitting diode lasers to be described in Sect. 12.12 of the last chapter. As the Bragg mirrors are fabricated inside the fiber, the operation of the laser would, therefore, be misalignment proof.

10.9.4   Scaling up the Fiber Laser Power: Problems and Remedies

Like any laser, power scaling of the fiber laser should be possible by increasing the active length. As explained below, this is not entirely true for a fiber laser as with increasing power, certain nonlinear processes become operative and eat away the intracavity power. As we know, the very long and thin geometry of fibers has granted this laser two major advantages, viz., easy removal of waste heat and inbuilt capability to operate on the lowest order transverse mode or TEM_{00} mode. However, as the fiber has a very narrow core, an increase in power also greatly raises the intra-core optical power density and, in turn, the associated electric field. With increasing power, the field can reach values high enough to set in various nonlinear processes that basically convert laser light into other wavelengths. Thus, any attempt to increase laser power by increasing the length of the fiber will be counterproductive as a part of the laser light will be lost to the inevitable nonlinear processes. In a nutshell, although a narrow fiber facilitates the removal of waste heat, it also has a detrimental effect on the power achievable from the fiber laser. The remedy of course will be to put a check on the occurrence of nonlinear phenomena by reducing the intracavity power density. This can be readily achieved by raising the diameter of the core, although at the expense of the spatial quality of the laser beam, as a wider core will support higher order transverse modes. The obvious question that arises here is whether the higher order modes can be suppressed in a wide aperture fiber laser to preserve the spatial quality of its emission. One popular technique has been to exploit the bending loss of a fiber to an advantage and subdue the higher order modes. As we know, light passes through a straight core without any spillage because it shines on the core-clad interface at angles of incidence greater than the critical angle (θ_c). However, when the fiber is bent, a part of the light will inevitably strike the interface at angles $< \theta_c$ and will escape as it travels down the core. This situation has been depicted in Fig. 10.27, which basically illustrates that the 90° complement Φ to the angle of incidence θ always increases as the light ray travels from the straight to bent region of an optical fiber. The ray A makes a large angle Φ_1 with the core-clad interface in the straight section of the fiber, while ray B makes a relatively smaller angle Φ_2 with this interface. Both these rays travel down the fiber, and as the fiber bends, they, respectively, now intersect the interface at angles Φ'_1 ($< \Phi_1$) and Φ'_2 ($<\Phi_2$). As Φ is complementary to θ, the angle of incidence, it is equivalent to saying that the angle of incidence of a ray always decreases as it travels from the straight to bent portion of a fiber. In the ray optics description, a higher order mode can be visualized as light rays travelling close to θ_c, the critical angle of incidence, while in the case of the TEM_{00} mode, the rays are always incident at

Fig. 10.27 The fact that the 90° complement (Φ) to the angle of incidence (θ) of a ray at the core-clad interface always increases as it travels from the straight to the bent zone of an optical fiber is illustrated in this figure

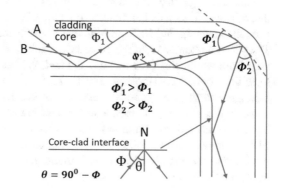

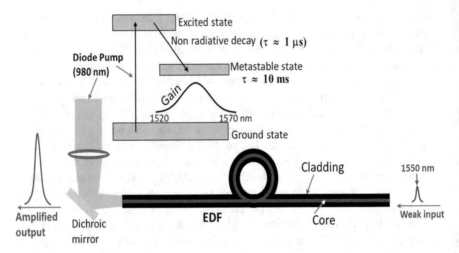

Fig. 10.28 Erbium-doped fiber (EDF) when pumped by a diode laser emitting on 980 nm acquires gain within ~1520 to 1570 nm range of wavelengths. The EDF amplifier therefore can amplify light at around 1550 nm wavelength

angles much higher than θ_c. Thus, rays A and B essentially signify the higher order and TEM$_{00}$ modes, respectively. It is therefore very likely that a part of light belonging to a higher order mode will leak out through the bent surface of the fiber and less likely that such a phenomenon will occur for the TEM$_{00}$ mode. Clearly therefore a higher order mode will experience greater bending loss than a lower order mode. The sharper the bending is, the greater this loss will be. By judicious selection of the sharpness of the bend, the higher order modes can be made so lossy that they simply die out, allowing oscillation on only the TEM$_{00}$ mode even for a wide aperture fiber. Power scaling in the operation of a fiber laser is therefore possible without compromising the spatial quality of its emission.

10.9.5   The Fiber Amplifier

Similar to any optical amplifier, an active fiber can also amplify light of frequency that falls within its gain bandwidth. The case of an erbium-doped fiber amplifier has been illustrated in Fig. 10.28. Erbium has a strong absorption at 980 nm wavelength. The 980 nm photons from the diode laser excite the erbium ions to a higher energy state from where they drop down to the metastable state and get stuck there, and, in turn, gain builds up across ~1520 to 1570 nm region of wavelengths. Photons belonging to the weak optical signal at around 1550 nm wavelength will induce stimulated emissions in these excited species in the metastable state. This will cause burning of the gain resulting in the amplification of the input signal. The ability of EDF to amplify light at ~1550 nm, where glass fibers are practically lossless, has profoundly impacted the fiber-optic communication systems and will be elaborated in the next volume.

Chapter 11
Molecular Gas Lasers

11.1 Introduction

In total contrast to atomic lasers where the lasing transition lies far above the ground energy level, in the case of the molecules, this can stay close to the ground state. A molecular gas laser derives at least two major advantages from the proximity of the lasing levels to the ground energy state: very high quantum efficiency and the ability to provide emission covering the enormous span of the infrared spectrum encompassing the near-, mid-, and far-infrareds. We need to take a closer look at the basics of molecular spectroscopy to appreciate these facts.

11.2 Basics of Molecular Spectroscopy

A molecule is basically made up of two or more atoms and can be thought of as a group of positively charged nuclei surrounded by a negatively charged electronic cloud. Like an atom, a molecule can also be electronically excited. However, upon supplying a quantity of energy that is much too small compared to that required to cause an electronic excitation, the atomic nuclei can also undergo vibrating motion within the nonrigid molecular framework[1] about the center of mass (CM). The molecular vibration is essentially a periodic motion of its constituent atoms, and the principle of conservation of linear momentum precludes any motion of the CM. Similar to the energy of the electrons, the energy of vibration is also quantized. Let us consider here the simplest case of a diatomic molecule, and the vibration of the two atomic nuclei about the CM is akin to the oscillation of two bodies joined by

---

[1] The intrinsic rigidity of a solid inhibits such vibrating motion of the atoms constituting a molecule as in case of a gas or liquid.

© The Author(s), under exclusive license to Springer Nature Switzerland AG 2023

D. J. Biswas, *A Beginner's Guide to Lasers and Their Applications, Part 1*,
Undergraduate Lecture Notes in Physics,
https://doi.org/10.1007/978-3-031-24330-1_11

Fig. 11.1 Oscillation of two masses m_a and m_b joined by a spring is equivalent to an oscillator of same spring constant but of reduced mass μ

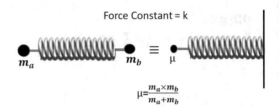

a spring (Fig. 11.1). Such a two-body oscillator of mass m_a and m_b is identical to a single oscillator of reduced mass[2] $\mu = \frac{m_a \times m_b}{m_a + m_b}$, and the frequency of oscillation ν_0 is given by

$$\nu_0 = \frac{1}{2\pi} \sqrt{(k/\mu)} \tag{11.1}$$

where k is the spring or force constant.

A quantum mechanical description of the vibration of the diatomic molecule reveals that vibrational energy E_v is quantized and is given by

$$E_v = \left(v + \frac{1}{2}\right) h\nu_0 \tag{11.2}$$

where v is the vibrational quantum number that can assume only integer values

$$v = 0, 1, 2, 3, \ldots\ldots\ldots\ldots \tag{11.3}$$

Upon substituting ν_0 from Eq. 11.1 into 11.2, the vibrational energy levels of a diatomic molecule are obtained as.

$$E_v = \left(v + \frac{1}{2}\right) \hbar \sqrt{(k/\mu)} \tag{11.4}$$

The spacing between two adjacent vibrational energy levels, viz., $E_{v+1} - E_v$, can be expressed from Eq. 11.2 as

$$E_{v+1} - E_v = \left(v + 1 + \frac{1}{2}\right) h\nu_0 - \left(v + \frac{1}{2}\right) h\nu_0 \tag{11.5}$$
$$= h\nu_0$$

This being independent of v, the molecular vibrational levels are equispaced like a stepladder. This, however, is true only for the low-lying vibrational levels. With

---

[2]Reduced mass is a hypothetical concept that greatly simplifies the analysis of a two-body problem by reducing it to that of a single body. For example, the vibrating or rotating motion of two bodies will be recreated by a single body if its mass were the product of the individual masses of the two bodies divided by their sum.

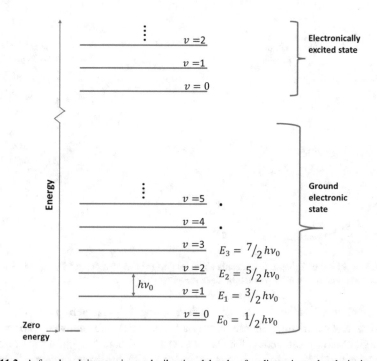

Fig. 11.2 A few low-lying equispaced vibrational levels of a diatomic molecule in its ground electronic state. To illustrate the enormity of energy required for an electronic excitation compared to that for vibrational excitation, the first electronic excited state is also shown for sake of comparison. In order to accommodate the electronic excitation in the same energy scale, the energy axis has been clipped appropriately

increasing v, the amplitude of vibration also increases, and soon the stretching and squeezing of the molecular bond becomes too large to obey Hooke's law[3] anymore. This manifests in the gradual reduction of the spacing between successive vibrational levels with increasing v and has a strong bearing in the operation and applications of molecular lasers; most notable among them is molecular laser isotope separation. Few low-lying vibrational levels of a typical diatomic molecule that is in its ground electronic state are shown in Fig. 11.2. The first electronic excitation of the molecule has also been indicated in the same figure to compare the energy of electronic excitation to that required to excite a quantum of molecular vibration. The energy axis of this figure is appropriately snipped to accommodate electronic excitation of the molecule that energy-wise can easily be two orders of magnitude more than that required for vibrational excitation. This essentially indicates that the vibrational energy levels that participate in the process of lasing reside very close to

---

[3]Named after its seventeenth-century discoverer, British physicist Robert Hooke (1635–1703), Hooke's law states that the deformation of an object is linearly proportional to the deforming force as long as the magnitude of deformation is small.

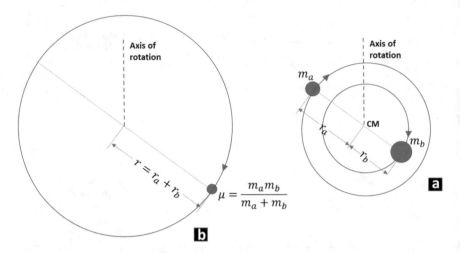

Fig. 11.3 Rotation of a diatomic molecule about its CM (**a**) can be recreated by the rotation of its reduced mass μ about an axis interatomic distance away (**b**)

the ground level unlike in the case when lasing originates from an electronic transition. Not surprisingly therefore, the molecular lasers are endowed with remarkably high quantum efficiency. The story doesn't end here though; a vibrating gas molecule can also rotate freely, and in compliance with the dictate of the microscopic world, the energy of rotation of the molecules too is quantized. We need to know here as to how the molecular rotation will influence the molecular spectra in general. For this, we once again consider a diatomic molecule and picture it as two rotating atoms of mass m_a and m_b separated by a distance r (Fig. 11.3a). The location of the center of mass of the molecule will obviously be governed by the equation

$$m_a \times r_a = m_b \times r_b \tag{11.6}$$

where r_a and r_b are the distances of atoms a and b, respectively, from the CM. The moment of inertia[4] of this molecule rotating about an axis passing through its CM and perpendicular to the line joining the two atoms can be expressed as

$$I = m_a r_a^2 + m_b r_b^2 \tag{11.7}$$

Combining Eqs. 11.6 and 11.7 we obtain

---

[4] The role that moment of inertia plays in the rotational motion is akin to what mass plays in case of a linear motion. More specifically, moment of inertia governs the torque required to produce a desired angular acceleration of a rigid body about its axis of rotation just as mass determines the force essential to produce a desired acceleration.

$$I = \frac{m_a m_b}{m_a + m_b}(r_a + r_b)^2$$
$$= \mu r^2 \tag{11.8}$$

where μ is the reduced mass of the molecule. The rotation of the diatomic molecule thus can equivalently be thought to be the rotation of a single mass μ about an axis located a distance r away, where r is the distance between the two atoms (Fig. 11.3b). The rotating molecule possesses an angular momentum, and the quantization of its rotational energy, which essentially follows from the quantization of angular momentum, can be expressed as[5]

$$E_J = \frac{J(J+1)h^2}{8\pi^2 I} \tag{11.9}$$

where J is the rotational quantum number. Upon substitution of B for $\frac{h^2}{8\pi^2 I}$, we obtain

$$E_J = BJ(J+1); \quad J = 0, 1, 2, 3 \tag{11.10}$$

B is denoted as the rotational constant of the molecule and is given by

$$B = \frac{h^2}{8\pi^2 I} \tag{11.11}$$

The spacing between two successive rotational levels can therefore be readily obtained as

$$E_{J+1} - E_J = B\{(J+1)(J+2) - J(J+1)\}$$
$$= 2B(J+1) \tag{11.12}$$

Unlike the case of vibrational energy levels, the rotational energy level gap therefore depends on the rotational quantum number J and increases with increasing J. This behavior is depicted in Fig. 11.4 for a few low-lying energy levels. It can be readily shown from Eqs. 11.5 and 11.12 that the typical energy involved to cause a vibrational excitation of a molecule is several orders of magnitude more than what is required to excite it rotationally. The energy required to impart a rotational excitation of a molecule is so small that in a typical gaseous sample at ambient temperature nearly all the molecules are, as a matter of fact, in one or another excited rotational states even though all of them are in the ground vibrational state.

---

[5] The rotation of a molecule sets in a centrifugal force that tends to pull the atoms apart. This, in turn, increases the moment of inertia of the molecule. With increasing J, this effect becomes appreciable and is accounted for by adding a centrifugal distortion correction term to the expression of rotational energy level as, $E_J = BJ(J+1) - DJ^2(J+1)^2$, D is known as the centrifugal distortion constant.

Fig. 11.4 The rotational energy levels are not equispaced, and the gap between any two consecutive levels increases with increasing J

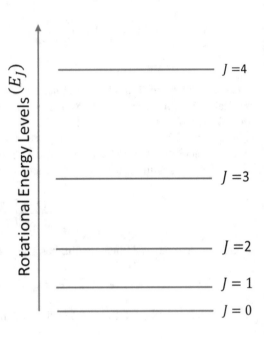

11.2.1 Vibration-Rotation Spectra

The smallness of the energy involved in molecular rotation allows us to conclude that the molecules, free to move in a gaseous environment, are almost always rotating regardless of their vibrational state. It is therefore not possible for a gaseous medium to emit light of pure vibrational origin. It is not surprising that pure vibrational spectra can be obtained only from liquids where interactions among the neighboring molecules inhibit their rotation. In the case of a gas, the gaps in the vibration energy ladder are essentially packed with rotational energy levels. A few low-lying rotational levels ($J = 0$, 1, 2, 3, and 4) in both the ground and first excited vibrational levels are shown in Fig. 11.5. Under ambient conditions, all the available molecules usually reside in the ground vibrational state and are distributed among the rotational levels following a Boltzmann distribution as shown in Fig. 11.6. The maximal behavior that the rotational population exhibits with energy originates from the degeneracy[6] of the rotational levels and also the fact that the degree of degeneracy rises with the rotational quantum number J. Optical transitions from a rotational level of one vibrational state to a rotational level belonging to another vibrational state are governed by the following set of spectroscopic selection rules:

---

[6] If two or more different physical states are located at the same energy level, then they are said to be degenerate. The number of such different states is called the degree of degeneracy.

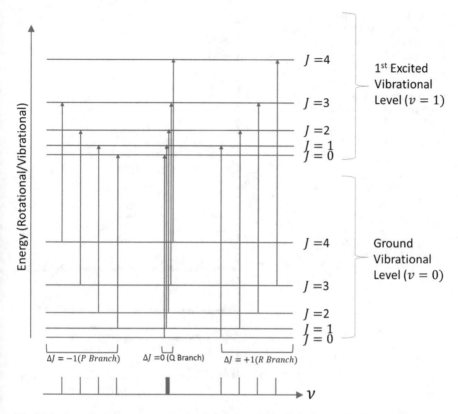

Fig. 11.5 A qualitative representation $J = 0$, 1, 2, 3, and 4 levels in the $v = 0$ and $v = 1$ vibrational levels along with P, Q, and R branch emission lines. As seen, the P and R branch transitions are spread over a range of frequencies, while all the transitions of the Q branch occur at the same frequency if the effect of centrifugal distortion is ignored. The Q branch line is therefore the most intense

$$\Delta v = \pm 1 \text{ and } \Delta J = 0 \text{ or} \pm 1 \tag{11.13}$$

This essentially means that if the vibrational and rotational quantum numbers of the originating energy level are v and J, respectively, then that for the terminal level can have either $v + 1$ or $v - 1$ as vibrational quantum numbers and $J-1$, J, or $J +1$ as the rotational quantum numbers. Such transitions between two rotational levels belonging to two different vibrational levels are termed ro-vibrational transitions. The transitions $J \rightarrow J - 1$ for which $\Delta J = -1$ are called P branch transitions, $J \rightarrow J$ for which $\Delta J = 0$ are called Q branch transitions, and $J \rightarrow J + 1$ for which $\Delta J = 1$ are called R branch transitions. As seen from Fig. 11.5, of these three branches, R branch lines have a higher frequency than P branch lines and Q branch lines lie in between. Furthermore, all the Q branch lines belonging to a particular vibrational transition have identical frequencies. In the typical ro-vibrational spectra, the Q branch therefore appears much stronger compared to the P and R branches wherein the emission

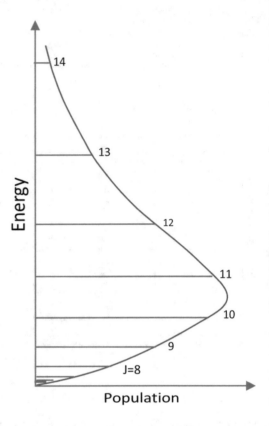

Fig. 11.6 The population of a vibrational level is distributed among the rotational levels satisfying Boltzmann distribution. The maximal behavior owes its origin primarily to the degeneracy of the rotational levels

lines are well separated from each other in frequency. The seemingly innumerable ro-vibrational transitions of frequencies spanning a wide infrared range of the e-m spectrum render tunability to the emission of molecular gas lasers.

11.3 Carbon Dioxide Lasers

Equipped with this elementary level of knowledge of molecular spectroscopy, we are now in a position to appreciate the working of a carbon dioxide laser, ubiquitous today as far as the industrial and medical applications are concerned. Invented in 1963 by an American-Indian scientist, Chandra Kumar Naranbhai Patel (b–1938), the CO_2 laser not only has the distinction of the very first molecular gas laser but was also the first to be used as a "light scalpel" to make bloodless incisions on the human body. Unlike most gas lasers, the CO_2 laser performs quite reliably across a wide range of operating pressures, allowing ready scalability of its power output at a remarkably high (10–20%) electro-optic efficiency. At low pressures, it usually operates in a continuous mode generating optical power that can vary from almost a watt to tens of thousands of watts. At an elevated gas pressure, on the other hand, this laser is mandatorily operated in pulse mode and capable of generating power to the tune of GW, albeit for a short duration. The coherent light that it generates can be discretely tuned over ~9 to ~11 μm wavelength with the strongest emission occurring near 10.6 μm. At these wavelengths, light exhibits excellent absorption by dielectric materials such as ceramics, biological tissues, water, and many more. This allows efficient transport of energy from the CO_2 laser beam to such substances. Consequently, a *cw* CO_2 laser, even with modest power (tens of watts), suffices for use as a surgical tool in a variety of medical applications as well as an ablation[7] tool for processing a host of dielectric materials. Metal, however, shows very high reflection at these wavelengths, and therefore, the usage of a high-power laser is mandatory for material processing applications in the case of metals. In addition to material processing applications of metals and nonmetals and medical applications, the other important areas that have greatly benefitted from the advent of the CO_2 laser include range finding, molecular laser isotope separation, and spectroscopy. By virtue of their high quantum efficiency, ability of low and high pressures as well as cw and pulse operations, high repetition rate capability, wide tunability, readily achievable spectral and spatial purity of emission, etc., CO_2 lasers have emerged today as one of the most popular and well-researched lasers.

11.3.1   Energy Level Representation of the Operation of a CO_2 Laser

Carbon dioxide is a symmetric linear triatomic molecule with the carbon atom at the center and the two oxygen atoms located symmetrically on its two sides. The constraint that the CM of a molecule remains stationary during vibrational motion allows the CO_2 molecule to vibrate in three distinct ways as depicted in Fig. 11.7. These three modes of vibrations are known as symmetric, bending, and asymmetric modes. In the symmetric mode of vibration, the carbon atom is always stationary, and the oxygen atoms move symmetrically toward it during one half cycle and away from it over the next half cycle. In the bending mode of vibration, all three atoms move perpendicular to the internuclear axis. In one half cycle, the carbon atom moves upward, while the oxygen atoms move downward to ensure that the molecular CM remains motionless. In the next cycle, the oscillating atoms, understandably, reverse their direction of motion. In the case of asymmetric mode of vibration, the symmetry of the molecule is lost. In one half cycle, one C-O bond is stretched, while the other is squeezed, and in the next half cycle, exactly the opposite happens to preserve the stationary condition of the CM. It goes without saying that the energies of vibration for all three modes, namely, symmetric, bending, and

---

[7]Laser ablation is a physical process wherein rapid deposition of optical energy on a material surface causes excessive localized heating of a thin layer and, in turn, its removal directly through vaporization.

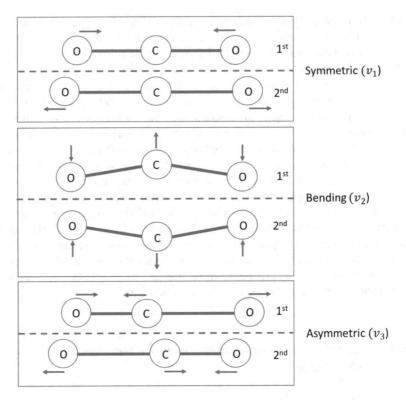

Fig. 11.7 Schematic representation of the three vibrational modes of a CO_2 molecule

asymmetric, are quantized and the respective quantum numbers are usually denoted as v_1, v_2, and v_3. The energy required to excite one quantum of vibration is least for the bending mode and highest for the asymmetric mode. The symmetric mode lies in between, and, as a matter of fact, one quantum of energy in the symmetric mode is nearly twice as large as that in the case of bending. Furthermore, these three modes of vibration are independent of each other, and, consequently, a CO_2 molecule can vibrate simultaneously on all three modes. An excited vibrational state is therefore denoted as $(v_1v_2v_3)$. This essentially means that the molecule in this state has v_1, v_2, and v_3 quanta of excitations in the respective vibrational states. For example, (100) represents an energy state with one quanta of excitation of the symmetric stretch mode and no excitation of either bending or asymmetric stretch modes, while (111) represents a state with one quanta of excitation of all three vibrational modes. The ground energy state with no vibrational excitation will therefore be described by (000). Few low-lying vibrational states that participate in the process of lasing are shown in Fig. 11.8. The first excited state of asymmetric vibration, namely, (001), serves as the upper laser level, while both the (100) and (020) levels act as the lower laser levels. The emissions in the 10 and 9 μm bands originate from the ro-vibrational transitions between the (001) and (100) and (001) and (020) levels,

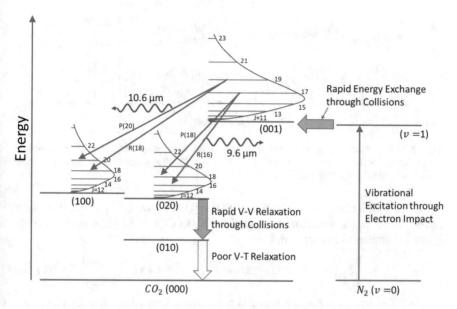

Fig. 11.8 A few vibrational energy levels, along with the rotational manifold of some of them, in the ground electronic state of the CO_2 molecules that participate in the process of lasing are schematically illustrated in this figure. The first vibrationally excited level of the N_2 molecules is also shown here

respectively. Selective transfer of the population from the ground to the upper laser level is possible in pure CO_2 gas through electron impact collisions in an electric discharge as shown below:

$$CO_2(000) + e(\text{KE}) \rightarrow CO_2(001) + e(\text{reduced KE}) \qquad (11.14)$$

This, however, is not very effective, as the excitation of the lower laser levels (100) and (020) through such collisions cannot be ruled out. Quite expectedly therefore the very first laser that made use of pure CO_2 gas produced output only at a minuscule level. Soon after operating the first CO_2 laser in 1963 at Bell Labs located in New Jersey, Patel demonstrated a remarkable improvement in the performance of this laser by adding N_2 gas in an appropriate quantity. N_2, being a diatomic molecule, has only one mode of vibration, and the fact that its first vibrationally excited state lies close to the (001) makes all the difference. When CO_2 and N_2 gases are mixed in equal proportions, nearly half of the collisions of the electrons will take place with N_2 molecules, causing their efficient transfer to their first vibrationally excited state.

$$N_2(v=0) + e(\text{KE}) \rightarrow N_2(v=1) + e(\text{reduced KE}) \qquad (11.15)$$

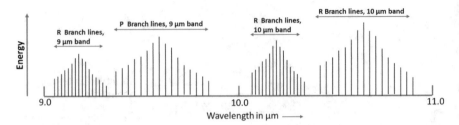

Fig. 11.9 A qualitative representation of energy on the rotational lines of the P and R branch transitions of 9 and 10 μm emission bands of a CO_2 laser. As seen, R branch lines are more closely spaced compared to the P branch lines

Owing to the close proximity of the $N_2(v = 1)$ and CO_2 (001) levels, extremely rapid transfer of the vibrational energy from N_2 to CO_2 takes place upon collision through resonant V-V relaxation.[8]

$$N_2(v = 1) + CO_2(000) \rightarrow N_2(v = 0) + CO_2(001) \qquad (11.16)$$

The slight energy mismatch between these two energy levels is readily compensated by the ever present thermal energy that prevails in the ambient conditions. The population that appears at the (001) level is, as we know, distributed among its rotational levels. Likewise, the vibrational population, mostly thermal,[9] present in the lower laser levels (100) and (020) is also distributed within their respective rotational levels. Ro-vibrational transitions between (001) and (100) and between (001) and (020) result in the possibility of lasing on numerous P and R branch lines originating from these transitions. These emission lines constitute the 10 and 9 μm bands across which a CO_2 laser can be tuned (Fig. 11.9). Certain spectroscopic rules forbid the presence of rotational levels with even quantum numbers in the (001) level and odd quantum numbers in both the (100) and (020) levels. This not only results in the absence of all odd rotational lines in the P and R branch transitions but also clearly eliminates the possibility of the occurrence of any Q branch transitions in the operation of a conventional CO_2 laser.

The onset of lasing gradually brings the population from the upper laser level to the corresponding lower laser level. Level (010) is located approximately halfway between the ground and either of the lower laser levels. The ensuing collisions between CO_2 molecules residing in the ground and lower laser levels result in near V-V relaxation of lower laser level population to the (010) level.

---

[8] In a gaseous mixture, a vibrationally excited molecule can transfer its excitation to a molecule of another gaseous species through a collision. Such an energy transfer process is called V-V relaxation. If the second molecule has a vibrational level closely matching that of the first, energy transfer process then occurs at an extremely fast rate and is called resonant V-V relaxation.

[9] Thermal population of an energy level is the population that intrinsically resides here following the Boltzmann distribution law at the ambient temperature of the medium.

$$CO_2(100) + CO_2(000) \rightarrow 2CO_2(010) \tag{11.17}$$

$$CO_2(020) + CO_2(000) \rightarrow 2CO_2(010) \tag{11.18}$$

The slight mismatch of energy on the left and right hand sides of both these equations is well compensated by the ever-present thermal energy making the relaxation of the population to (010) an extremely rapid process. The radiative half-life of (010) is very high, and the population from this level can return to the ground state only through collisions. As vibrational energy needs to be realized here as translational energy of the colliding body, such a relaxation process, termed V-T relaxation, intrinsically is a rather slow process. Helium, however, has been found to be more effective in removing this vibrational energy from the CO_2 molecule and sending it back to the ground state upon colliding with it.

$$CO_2(010) + He(\text{KE}) \rightarrow CO_2(000) + He(\text{elevated KE}) \tag{11.19}$$

Helium obviously carries this vibrational energy as its KE. The addition of helium in large quantities with CO_2 and N_2 gases therefore not only greatly improves the performance of a CO_2 laser but also facilitates its *cw* operation, as it helps the CO_2 molecules to readily return to the ground state and continue to participate in the process of lasing. Helium, as we shall find below, also helps stabilize the discharge of a high-pressure CO_2 laser that is intrinsically very prone to arcing. As a matter of fact, in the conventional operation of a CO_2 laser, helium constitutes the major part of the gas mixture. The use of CO_2, N_2, and helium in the proportion of 1:1:8 is not uncommon in the typical operation of these lasers. It is appropriate here to elaborate a little on the popular schemes to pump this laser and on its emission characteristics.

11.3.2   Low-Pressure CO_2 Lasers, Longitudinal Pumping, and cw Operation

The CO_2 laser, like most other gas lasers, is also electrically pumped by taking advantage of a gas discharge. As we know from Chap. 7, at low gas pressures, the breakdown voltage is also low, and therefore discharge can be effected by maintaining a longitudinal geometry. This greatly simplifies the operation as a longitudinal discharge readily fills a major part of the gaseous mixture allowing homogeneous interactions between the electrons and all the molecules/atoms present in the mixture. This, in turn, results in the realization of uniform gain across the length and breadth of the active medium. Furthermore, the *cw* operation capability is intrinsic to the low-pressure CO_2 laser as a glow discharge can be sustained here as long as the electrical pumping network continues to feed it with energy at the required rate. Such lasers, when suitably designed, are capable of producing output power that can vary widely from approximately a watt to tens of thousands of watts. To gain insight into the diversity of the operation of the *cw* CO_2 laser, we focus our

Fig. 11.10 Schematic
illustration of the operation
of a simple water-cooled cw
CO_2 laser along with its
associated electrical
pumping unit

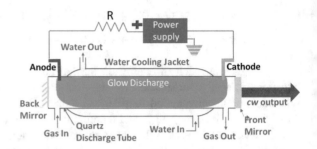

attention on a basic device, the kind used by its inventor, C. K. N. Patel, in 1963. This is schematically illustrated in Fig. 11.10 which essentially is a variant of Fig. 7.17 that was made use of in Chap. 7 to introduce the general features of an electrically pumped *cw* gas laser. CO_2, N_2, and He gases, mixed in the right proportion, flow in and out through the inner tube of a double walled quartz enclosure, while water flows through its outer jacket to help remove the heat that stays back in the system following lasing. The flow rate of the gas mixture is adjusted to maintain the required operating pressure (a few torrs to a few tens of torrs). Gain is established by striking a discharge longitudinally between the two electrodes placed at the two ends of the inner tube. The high quantum efficiency in conjunction with the efficient roles played by N_2 and He in the creation and sustenance of population inversion allows the CO_2 lasers to convert a sizeable fraction, not feasible for atomic lasers, of the input power into coherent light. This makes it possible for this laser of a typical active length of 1 m to produce several tens of watts of optical power quite reliably employing the rudimentary method of removal of residual heat from the lasing medium as shown in Fig. 11.10. However, scaling up the power output of the laser is not possible this way as the removal of heat by simply flowing water over the outer surface of the quartz tube will no longer be adequate. The construction of a high-power laser would be more involved, as the hot gas would be required to be removed from the discharge tube, cooled and reused. The recirculation of the gas is possible by flowing it longitudinally along the length of the discharge or transversely across the discharge, and the same is described in the following sections.

11.3.2.1 Power Scaling of *cw* CO_2 Lasers with Longitudinal Gas Flow

If the laser has to deliver higher power, the electrical input to the system should also be accordingly higher. This essentially means that more residual heat would be required to be removed from the lasing medium. Let us consider a situation when the *cw* power of the laser is to be scaled up to the kilowatt level. Even at an electro-optic efficiency as high as 10%, 10 kW electrical power is required to be transferred to the discharge if the laser has to generate coherent light of a kW power. The residual power of 9 kW is much too large to be carried away by the water flowing over the discharge tube, an arrangement known to be adequate for low power *cw* CO_2 lasers. The temperature of the gas rises so rapidly here that the hot gas has to be mandatorily

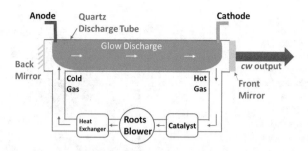

Fig. 11.11 The schematic illustration of the laser head and gas recirculatory loop of a longitudinal flow high-power cw CO_2 laser. The hot gas after its passage through a catalyst bed for recombining CO and O back into CO_2 flows through a heat exchanger before returning back to the laser head

flown in and out of a heavy-duty heat exchanger if its temperature has to be brought down and maintained at an acceptable level. This essentially means that the laser head needs to be integrated with a gas recirculation loop comprising an appropriate gas blower and a heat exchanger. A very simple arrangement where the hot gas flows longitudinally along the length of the discharge is depicted in an extremely qualitative manner in Fig. 11.11. The rate at which the gas must flow through the recirculatory loop should be commensurate to the rate at which heat is being dissipated in the lasing medium.

The requirement of rapid flow of gas through the narrow and long discharge tube calls for a gas blower that is inherently capable of providing large volumetric flow at a high-pressure drop.[10] Specially designed roots or turbo blowers meet these requirements and normally find application here. The exorbitantly high cost of such heavy-duty blowers, however, limits their use up to approximately kW class of lasers. It should be noted here that in the discharge, a fraction of CO_2 molecules also dissociate into carbon monoxide and oxygen upon electronic impact.

$$CO_2 + e(\text{KE}) \rightarrow CO + O + e(\text{Less KE}) \tag{11.20}$$

In a flowing gas laser system, these dissociated products do not accumulate and hence pose no problem in the operation of the laser. In a sealed off system, on the other hand, they build up in time and detrimentally affect the performance of the laser. In the sealed off operation, carbon monoxide and oxygen are often catalytically recombined back into CO_2 to ensure reliable operation of the laser. The gas recirculation loop, as seen in this figure, is therefore also equipped with a catalytic re-converter. The hot and CO_2 depleted gas flows out of the front end of the laser head and passes successively through the catalyst and heat exchanger before the cold and CO_2 replenished gas flows into the laser head from the rear end.

---

[10]Pressure drop is a measure of the hindrance that a fluid experiences while flowing through a narrow and long pipe and is akin to the resistance that a conductor poses to the flow of electric current through it.

Fig. 11.12 An oversimplified cross-sectional view of the laser head and the transverse gas flow-based recirculatory loop. The longitudinal slits on either side of the discharge loop allow comfortable passage of the gas across the discharge

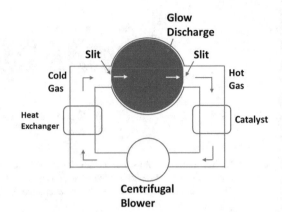

11.3.2.2   Power Scaling of *cw* CO₂ Lasers with Transverse Gas Flow

For the construction of more powerful multikilowatt (~10 kW or more) CO_2 lasers, the gas flow geometry is normally changed from longitudinal to transverse, which dramatically reduces the pressure drop for the flow of gas. Large volumetric gas flow in this case is possible by the readily available and less expensive axial or centrifugal gas blowers. The laser head is required to have two rectangular slits running along its length on either side to make it compatible with transverse flow. An oversimplified cross-sectional view of the laser head along with the transverse flow loop for gas recirculation is shown in Fig. 11.12. Commercial multikilowatt *cw* CO_2 lasers are normally based on such a transverse flow geometry that intrinsically offers high volumetric gas flow with relative ease.

11.3.3   High-Pressure CO₂ Lasers, Transverse Pumping, and Pulsed Operation

We have seen that the output power that a *cw* CO_2 laser is capable of delivering can be raised by appropriately augmenting the pump power followed by a remedial measure to remove the additional heat that is realized in the lasing medium as a consequence. There exists a major hurdle in scaling up the power of a *cw* CO_2 laser by raising its operating pressure beyond a point. The fact that a gaseous discharge always begins as a glow discharge but eventually grows into an arc within a time inversely proportional to the gas pressure[11] allows its *cw* operation up to about a few tens of torr pressure. Beyond this pressure, the discharge begins to constrict eventually culminating into an arc. The flow of current between the electrodes now

---

[11] Glow to arc transition time for a typical CO_2 laser gas mixture at atmospheric pressure is several hundreds of nanoseconds.

Fig. 11.13 At low operating pressures, the discharge occurs in a glow mode filling the majority of the accessible volume of the discharge housing (**a**). At a higher pressure, the discharge becomes constricted, and consequently current density becomes remarkably high. The current flows through a minimum impedance path and may exhibit many twists and turns (**b**)

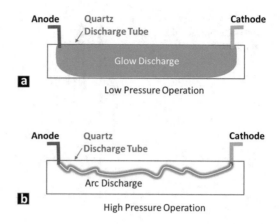

occurs through a path that offers minimum impedance often taking a zigzag course (Fig. 11.13). In an arc discharge, the pump energy is essentially dissipated into a greatly reduced gas volume causing an abrupt termination of the process of lasing. This renders the *cw* operation of a CO_2 laser impractical at higher gas pressures. As a glow discharge takes a finite time before it can grow into an arc, the high-pressure operation of the laser is possible only in a pulsed mode. However, effecting a longitudinal discharge at high pressure is not feasible as the breakdown voltage of a gap increases with pressure. The advent of pulsed transverse discharge pumping offers a practical solution to this problem. Realized first in the operation of CO_2 lasers, and subsequently applied to many more gas lasers, the transverse pumping geometry allowed scaling up of the operating pressure quite remarkably and, in turn, the achievable power from the laser. As described in Sect. 7.5.2.2, arc free discharge can be struck between two uniform field electrodes, causing homogeneous pumping of the interelectrode gaseous volume. Popular as transverse electric atmosphere (and abbreviated as TEA) CO_2 laser, it can in the single shot operation yield multi-joule energy in a pulse in the conventional operation with CO_2, N_2, and He in the proportion of approximately 1:1:8. The temporal shape of a typical TEA-CO_2 laser pulse is illustrated in Fig. 11.14 which exhibits a gain switched peak of width about 100 nanoseconds, followed by a longtail of several microseconds. The vibrational level of N_2, which has a near resonance with the upper laser level, is a metastable state. This can therefore hold the excitation energy for a long time and continues to transfer this energy to CO_2 even long after the process of lasing has begun through collisions and consequently contributes to the formation of the longtail. The peak power of the pulse from a conventional laser can therefore easily run into tens of MW to approximately 100 MW. The pulse compression techniques, described in the preceding chapter, can be readily applied to further boost the intra-pulse output power of a TEA-CO_2 laser.

The scaling up of the average power of the laser calls for its repetitive operation. The Spark gap that is usually used to switch the stored energy into the discharge in

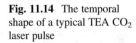

Fig. 11.14 The temporal shape of a typical TEA CO_2 laser pulse

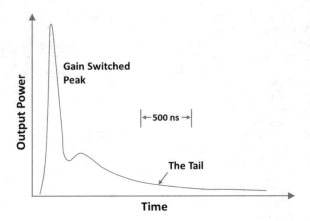

the single shot mode of operation (refer to Chap. 7, if you are hazy on this) is normally replaced by a thyratron[12] for repetitive switching of the TEA-CO_2 laser. The output power of the laser can be increased by increasing the repetition rate provided adequate measures are enforced to prevent any accumulation of heat in the lasing medium. As in the case of the operation of a high-power *cw* CO_2 laser, convective cooling of the hot gas is also possible here too by flowing it longitudinally or transversely. Incorporation of a catalytic re-converter is also mandatory, as before, to ensure reliable enduring performance of a high-power TEA-CO_2 laser.

11.3.4   Spectral Features of CO_2 Laser Emission

As we now know, the CO_2 lasers are capable of operating across a wide range of pressures that can vary from a few torrs to atmospheric pressure and even beyond. This results in a wide variation in the degree and nature of the gain broadening and consequently impacts the spectral characteristics of the emission of a CO_2 laser in a remarkable manner. To compare the spectral purity of the emission of the laser for low (~10 mbar) and high (~1000 mbar) pressure operations, we consider the laser to have a typical cavity length of 1 m. At 10 mbar pressure, both pressure and Doppler broadening are nearly same, and the net broadening of the gain, that is, convolution of the 2, equals ~100 MHz. The *FSR* of a 1 m cavity is 150 MHz. It is therefore equivalent to saying that the FSR of the laser cavity normally exceeds the lasing bandwidth or alternately a low-pressure CO_2 laser is expected to always lase on a single longitudinal mode (SLM). With increasing pressure, the homogeneous component of gain broadening begins to play a dominating role. At atmospheric pressure

---

[12]Thyratron is a variant of the vacuum valve that offers very fast switching of high voltage gas discharge and capable of handling remarkably high average power making it ideally suited as a switch for high repetition operation of TE gas lasers.

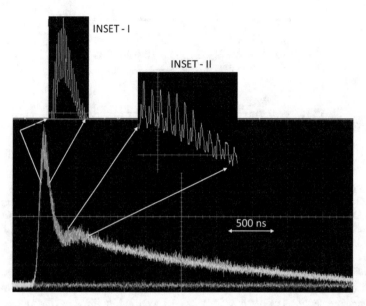

Fig. 11.15 The beating of the longitudinal modes manifests as temporal modulation of intensity in the pulsed output of a TEA-CO$_2$ laser. This temporal profile of the laser pulse was captured in the CO$_2$ laser laboratory of the Bhabha Atomic Research Centre, Mumbai

(1000 mbar) operation, typical for a TEA-CO$_2$ laser, when the pressure broadening can be as high as 5 GHz, gain on multiple longitudinal modes exceeds loss and can lase by exploiting the phenomenon of spatial hole burning. The deep modulation of intensity on the temporal profile of the pulse (Fig. 11.15) emitted from such a laser, a clear signature of mode beating, bears testimony to this fact. The modulation is also shown on a time expanded scale in the insets of this figure; inset-I displays a part of the gain switched peak, while inset-II depicts a small part of the pulse tail. Taking a closure look at this figure, we can readily identify the period of intensity modulation to be originating from the beating of two adjacent longitudinal modes. The irregularity in the pulsation also bears testimony to the free running condition of the laser wherein the modes are not locked. Clearly, therefore, a low-pressure CO$_2$ laser always gives out light much purer in color compared to when it operates at a higher pressure. The production of light of higher intensity by a *TEA*-CO$_2$ laser thus occurs at the expense of the purity of its color. Raising the purity of the color of light emitted by this laser, a prerequisite for many of its applications, has been extensively researched over the years.

11.3.4.1   SLM Emission from a Hybrid CO$_2$ Laser

Hybridizing low- and high-pressure gain cells in the same optical cavity presents an ingenious technique to effect SLM lasing in a TEA-CO$_2$ laser. A hybrid CO$_2$ laser is schematically described in Fig. 11.16, and the underlying physics operative here for

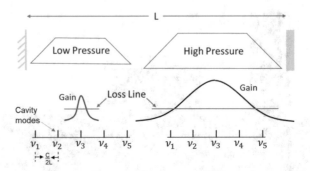

Fig. 11.16 In a hybrid CO_2 laser, a low- and a high-pressure gain cells are placed one after the other within the same optical cavity. As seen, only one cavity mode of frequency ν_3 will experience gain in both high- and low-pressure sections and therefore will be able to lase. A hybrid CO_2 laser will always exhibit SLM operation

SLM lasing is described below. Five adjacent cavity modes, namely, ν_1, ν_2, ν_3, ν_4, and ν_5, have been indicated with respect to the broadened gain of both the low- and high-pressure sections of the hybrid laser. As seen, while three cavity modes ν_2, ν_3, and ν_4 will experience net gain in the high-pressure section, only the lone mode of frequency ν_3 will have gain exceeding loss in the low-pressure section. The cavity mode with frequency ν_3 that experiences gain in both sections will therefore be able to lase. A hybrid laser will therefore always lase on *SLM*.

11.3.4.2 SLM Emission from a Ring Cavity CO_2 Laser

We know from Chap. 8 (Sect. 8.4) that a homogeneously broadened laser will always lase on only one mode possessing the highest gain in a unidirectional ring cavity laser. Therefore, if a TEA-CO_2 gain medium is placed in a unidirectional ring cavity, lasing will be restricted on SLM. The use of a small cell containing SF_6 gas of suitable pressure at an appropriate location inside a ring cavity can force unidirectional lasing in a TEA-CO_2 laser and, in turn, its SLM operation. A simple ring cavity containing the TEA-CO_2 gain and SF_6 absorption cells and comprising only three mirrors, two fully reflective and one partially reflective output mirrors, is shown in Fig. 11.17. SF_6 gas essentially performs here the role of a saturable absorber. In the absence of any intracavity saturable absorber, as the gain medium is energized, the co- and counterpropagating waves grow equally giving rise to the standing wave formation resulting in bidirectional lasing. The standing wave causes the occurrence of spatial hole burning making the multimode lasing of the TEA-CO_2 laser inevitable even though its broadening of gain is homogeneous at the atmospheric pressure of operation. In the presence of SF_6 gas that absorbs CO_2 laser emission, the wave $M_2M_1M_3$ that enters the absorption cell upon experiencing transmission loss on mirror M_3 is weaker compared to the wave $M_3M_1M_2$ entering the cell from the opposite direction and is unable to saturate the absorption leading to

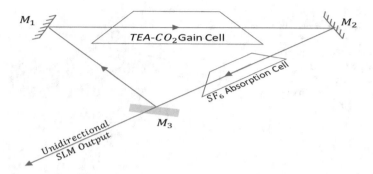

Fig. 11.17 An appropriately located saturable absorber inside a simple ring cavity can impose unidirectional emission and, in turn, SLM operation of a TEA-CO$_2$ laser

its annihilation. This prevents the formation of a standing wave and, in turn, elimination of spatial hole burning. The end result is thus the SLM operation from the TEA-CO$_2$ laser.

11.3.4.3 Tunable Emission from a CO$_2$ Laser

We know that the vibrationally excited CO$_2$ molecules are capable of emitting on a host of discrete ro-vibrational transitions spanning over the 9 and 10 μm bands of wavelengths. By making use of a dispersive resonator cavity, the emission of both *cw* and pulsed lasers can be tuned discretely across this wide range of mid-infrared wavelengths. However, before describing the tunable operation of a CO$_2$ laser, it is imperative to gain insight into the working of a dispersive optical cavity.

11.3.4.3.1 A Prism or a Grating as a Dispersive Element

We have learned from Chap. 2 that dispersion of light is the process of spatial splitting of its constituent wavelengths and occurs when it undergoes either refraction or diffraction. A prism is the most common optical element that disperses light through refraction. A grating, which basically comprises multiple tiny periodic grooves, is, on the other hand, an example of an optical element that disperses light through diffraction. The splitting of white light into its constituent colors by both a prism and a reflection grating[13] is qualitatively shown in Fig. 11.18. The resolution of a grating can be one to two orders of magnitude higher than that of a prism and therefore produces light with better spectral purity. It is not surprising that

---

[13] Although a grating can be both of reflection and transmission types, we would restrict ourselves to reflection gratings only as they normally find application in the operation of a tunable laser. A transmission grating is made up of multiple periodic slits. The light upon passing through these slits undergoes diffraction and, in turn, dispersion.

Fig. 11.18 Schematic
illustration of the dispersion
of light by a prism (top) and
a reflection grating (bottom).
The grating has a higher
resolving power and causes
wider spatial separation of
the wavelengths

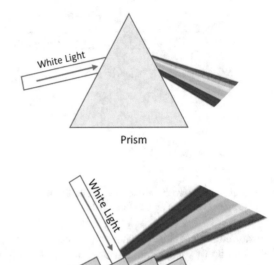

a grating usually finds application as the dispersive element in the operation of a
tunable laser. The operation of a laser with the prism as the wavelength tuner is pretty
straightforward and has been touched upon earlier in Chap. 6 (Sect. 6.5). As will be
seen below, the use of a grating as a wavelength tuner is a little trickier because
grooves with a unique geometry can only impart tunability to the emission of a laser.

11.3.4.3.2 Inability of a Conventional Grating to Perform as a Laser Cavity Tuning Element

We know from Chap. 3 that light of a given wavelength is diffracted into various
orders identified as zeroth, first, second, and so on. The zeroth order is nothing but
the normal reflection and understandably does not exhibit any dispersion, and the
dispersion increases with the order of diffraction. The diffraction of an incoming
beam of light comprising two wavelengths λ_1 and λ_2 as it is incident on a reflection
grating is schematically illustrated in Fig. 11.19. The diffraction of light off a grating
is described by the following grating equation:

$$d(\sin \alpha + \sin \beta_m) = m\lambda \tag{11.21}$$

where d is the periodicity of the grooves and is also called the grating element,[14] α is
the angle of incidence, m is the order of diffraction, and β_m is the corresponding

---

[14] As we know from the theory of diffraction, the grating element d must be close to the wavelength
of light to make the condition for its diffraction on the grooves favorable.

Fig. 11.19 The diffraction of a beam of light comprising of wavelengths λ_1 and λ_2 off a typical reflection grating. An enlarged view of the grating revealing the groove geometry is also shown here

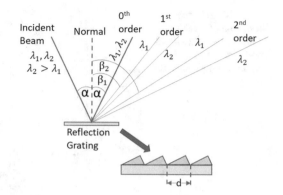

diffraction angle. The incident light, as seen, is diffracted into multiple orders. Clearly, the zeroth order does not show any dispersion as both λ_1 and λ_2 emerge in the same direction, and the dispersion increases with increasing order. As the intensity of the incident light at a given wavelength is distributed into different orders of the diffracted light, a conventional grating cannot be used as a wavelength tuner in the operation of a laser.

11.3.4.3.3   A Blazed Grating: An Ideal Laser Cavity Tuning Element

This limitation can be overcome by fabricating a grating specifically to yield the maximum diffraction efficiency on a particular order while minimizing the loss of power to all the remaining orders, in particular the zeroth. Such a grating is called a blazed grating and can be used as a tuning element in the operation of a laser both in the Littrow and non-Littrow modes. As shown in Fig. 11.20, in the Littrow configuration, wherein the angle of incidence equals the angle of diffraction, the grating performs the dual role of a tuner and feedback mirror, while in the non-Littrow mode, it is to be used in conjunction with a feedback mirror. The principle of operation of a blazed grating can be qualitatively understood by referring to Fig. 11.21 which is specifically drawn for first order blazing in the Littrow configuration. The blazing angle θ, which is essentially the angle between the surface parallel and the surface structure, is also the angle between the surface and facet or groove normal. Therefore, if the incident light strikes the grating along its facet normal, the diffracted ray will travel back in the same direction. If the two grooves or facets are spaced by a length $\lambda/2$, then the path difference of the two rays diffracted from the adjacent structures is clearly $\lambda/2 + \lambda/2$ or λ. If the incident light has a wavelength of λ, then these two rays would constructively interfere, or alternatively it can be stated that the grating would offer maximum diffraction efficiency in the first order for wavelength λ. The diffraction efficiency will progressively drop on either side of λ. It goes without saying that a grating blazed for wavelength λ in the

Fig. 11.20 Schematic illustration of a blazed grating in the Littrow and non-Littrow modes. The grooves are not shown for clarity

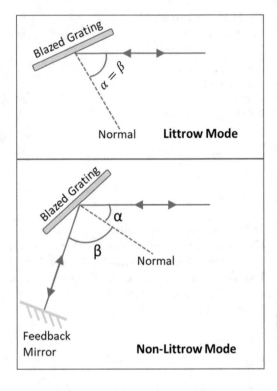

Fig. 11.21 The Littrow mode operation of a grating that is blazed for wavelength λ in the first order. The inset of the figure shows some relevant surfaces and normal

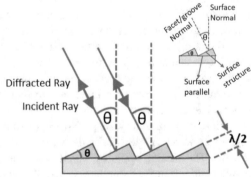

first order is inevitably also blazed for $\lambda/2$ in the second order, $\lambda/3$ in the third order, and so on. Furthermore, the resolving power R of a grating is a measure of its ability to spatially separate two closely spaced wavelengths and is given by

$$R = \frac{\lambda}{\Delta\lambda} = mN \qquad (11.22)$$

Fig. 11.22 The cavity arrangement of a typical grating tuned CO_2 laser

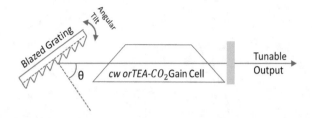

where $\Delta\lambda$ is the resolvable wavelength around the central wavelength λ, m is the order of diffraction, and N is the total number of grooves of the grating illuminated by the light beam. Thus, the resolution of the grating increases with diffraction order and the number of grooves which participate in the process of diffraction.

11.3.4.3.4   A Blazed Grating Tuned Operation of a CO_2 Laser

The tunable operation of a CO_2 laser with such a blazed grating in the Littrow configuration is illustrated in Fig. 11.22. Angular tilt of the grating will result in feedback and, in turn, lasing on varying wavelengths. Thus, the laser can be tuned, in principle, to all its ro-vibrational transitions depicted in Fig. 11.9 simply by varying the angle of incidence θ to the grating. The diffraction efficiency will obviously be highest at the wavelength for which the grating is blazed. There exist many more lasers with gain either discretely or continuously spread over a wide range of wavelengths. Needless to say, tunability can be imparted to the emission of these lasers by choosing a grating of appropriate blazed wavelength and groove periodicity. Furthermore, the resolving power of a grating can be increased by exposing a larger surface area of the grating surface to the beam of light. As we have seen in the operation of a dye laser (Sect. 10.3.1.2), this can be achieved by introducing a beam expander between the grating and the active medium.

Chapter 12
Semiconductor Lasers

12.1 Introduction

That certain materials behave like a semiconductor is believed to have been first scientifically documented in the work of British scientist Michael Faraday (1791–1867). He discovered in 1833 that the resistance of silver sulfide decreased with increasing temperature in complete contrast to the behavior of metals.[1] These semiconductor materials behave much like insulators at low temperatures but conduct electricity like metals at room or elevated temperatures. It, however, took more than a century before the electrical properties of a semiconductor could be manipulated by American engineer Russell Shoemaker Ohl (1898–1987) to build the p-n junction, which is now ubiquitous in electronic and optoelectronic devices. The p-n junction when suitably biased behaves like a diode and is regarded as the very first semiconductor electronic device. The transistor, a variant of the p-n junction, brought a revolution in electronics and, no wonder, fetched the 1958 Nobel Prize in Physics to its 1948 inventors William Bradford Shockley (1910–1989), John Bardeen (1908–1991), and Walter Houser Brattain (1902–1987). Bardeen incidentally became the only person to date to get the Nobel Prize in Physics twice; in 1972, he and his collaborators Leon N Cooper and John Robert Schrieffer were awarded the coveted prize for developing the BCS theory to explain conventional superconductivity. The credit of inventing the light emitting diode (LED) goes to Nick Holonyak[2] (1928–2022), an American engineer, who, in 1962, while working with General Electric, succeeded in generating spontaneous red light from a forward

---

[1] The resistivity of metals increases with temperature.

[2] In 2006, the American Institute of Physics announced five most important papers in the journals published by it since it was founded 75 years ago. One of these five papers happened to be the *Applied Physics Letters* paper that Holonyak coauthored with S. F. Bevacqua to report the creation of the first LED in 1962.

© The Author(s), under exclusive license to Springer Nature Switzerland AG 2023
D. J. Biswas, *A Beginner's Guide to Lasers and Their Applications, Part 1*,
Undergraduate Lecture Notes in Physics,
https://doi.org/10.1007/978-3-031-24330-1_12

biased p-n junction of GaAs semiconductor. This operation of LED provided a major impetus to build a semiconductor diode laser, and by the end of that year, Robert N. Hall (1919–2016), an American physicist, while working at General Electric, Schenectady, New York, generated coherent light from a gallium arsenide diode. The race to build the diode laser was so intense that almost simultaneously the groups working at the MIT Lincoln Laboratory, Massachusetts, and the IBM T.J. Watson Research Center also reported lasing from GaAs diode. Erstwhile Soviet Union was not behind either as in early 1963 Nikolay Basov and his associates developed a gallium arsenide diode laser at the Lebedev Physical Institute, Moscow. That in the following year he was awarded the Nobel Prize in Physics for his pioneering contribution to the invention of laser is now a history. The inbuilt tiny size of the semiconductor laser enabling its ready integration with other devices primarily led to an explosion of research activity across the world to perfect the diode laser technology by exploiting the unique properties that a semiconductor is known to offer. This has led to the emergence of diode lasers as the undisputed leader in the world of lasers; the readers may refer to the second volume of this book to catch a glimpse on the seemingly endless applications of these lasers covering almost all branches of science and technology. As a matter of fact, to say that today diode lasers have made profound impact on our day-to-day life will not be an overstatement; they are present everywhere – CD players, DVD players, barcode readers, laser pointers, cellphones, HD TV – and most importantly they are central to the ever-expanding global communication network making possible the hassle-free long distance trans-pacific and transatlantic video/audio calls.

To be able to appreciate the workings of diode lasers and their diversity and versatility, it is imperative to gain insight into the physics of semiconductors. An intuitive approach will be followed here to elucidate the basics of semiconductors prerequisite for an easy understanding of the working of all variants of diode lasers.

12.2   Basics of Semiconductor Physics

The conductivity of a material is essentially governed by the mobility of its constituent electrons, in particular the ones that are farthest from the nucleus of the respective atoms. It is worthwhile in this context to take another look at Bohr's atomic theory, introduced in Chap. 3, which pictures an atom as its electrons revolving around the positively charged nucleus in specific orbits with rapidly increasing radii. Electrons residing in the outermost orbits are the most loosely bound as they are attracted by the nucleus with the least force. To determine the energy levels of such electrons of the isolated atoms when they are brought closer and closer to form a solid, we consider the simplest case of two hydrogen atoms separated by a distance of R (Fig. 12.1). As long as R far exceeds $2r_0$ where r_0 is the radius of the ground state Bohr orbit, there is no appreciable overlap of the wave

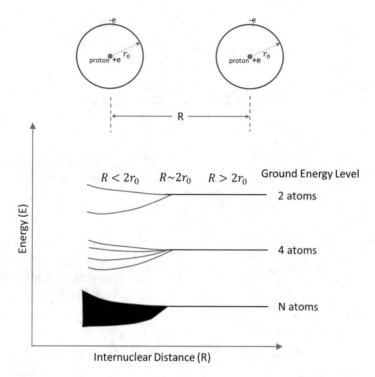

Fig. 12.1 Energy level as a function of interatomic distance. When two hydrogen atoms come close together, the ground state energy level splits into 2 due to the overlap of the corresponding electronic wave functions. Generally speaking, four atoms would split the original level into 4. In reality, there are countless atoms in a solid that would essentially form an energy band

functions[3] of the two atoms, and consequently the discrete ground energy level remains the same as that of the isolated atoms. However, as R approaches $2r_0$, the wave functions begin to overlap, and each electron orbits around both nuclei. Clearly, the wave function of each electron can be approximated as the combination of the wave function of the isolated atoms with both the same and opposite signs. The situation is represented in the illustrations of Fig. 12.2. The two wave functions essentially correspond to two different energy values. This is equivalent to saying that when the two hydrogen atoms are close enough, the ground state energy level splits into two. If there were four atoms instead of two, the energy level would then split into four as the number of new levels equals the number of interacting atoms. Generally speaking, if N atoms are brought together, the energy level would split into N separate levels. In a solid, an exceptionally large number of atoms are packed in a tiny volume, and consequently the levels come so close to each other that they essentially form an energy band, as seen in Fig. 12.1. It needs to be emphasized here that the formation of such energy bands is possible primarily for valence electrons

---

[3] The description of wave function is a little elaborative and is appended at the end of this chapter.

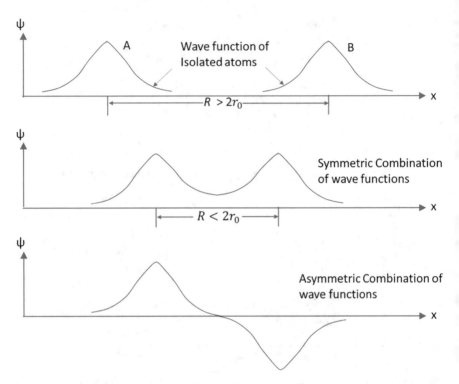

Fig. 12.2 Electronic wave function of two isolated hydrogen atoms (top trace). For small internuclear separation R, the isolated wave functions combine with both same (middle trace) and opposite (lower trace) signs

that are relatively loosely bound to the nucleus as they reside in the outermost shell. Clearly, wave functions of the inner electrons belonging to different atoms forming a solid exhibit very little overlap. We know from Chap. 4 that the shells are denoted by the principal quantum number n. It may be noted here that a shell with a principal quantum number n has an equal number of subshells denoted by orbital quantum number l that can possess integer values 0, 1, 2, 3, ...$(n-1)$. The maximum number of electrons that can reside in a subshell of orbital quantum number l is $2(2l + 1)$. Furthermore, the subshells with $l = 0, 1, 2, 3 \ldots$ are, respectively, denoted as s, p, d, f, etc. Therefore, the maximum occupancy of electrons in $s, p, d, f \ldots$ subshells are, respectively, 2, 6, 10, 14, and so on. In order to appreciate the bearing of the formation of the energy band by the valence electrons on the conductivity of a solid, we consider the typical case of a sodium atom that has an atomic number of 11. Each sodium atom thus has 11 electrons, and their distribution among the various shells and subshells is shown in Fig. 12.3. As seen, the shells with principal quantum numbers 1 and 2 are full, and the outermost subshell ($n = 3, l = 0$ and accordingly denoted as $3s$) is only half full, as only one electron resides here, while the occupancy of the s-subshell is two. This being the valence electron, when sodium atoms get closer to form a solid, the $3s$ energy level will be broadened to form an

Fig. 12.3 Of the
11 electrons that an atom of
sodium possesses, 10 fill all
the available slots in the first
and second shells, and the
11th electron therefore
resides in the $l = 0$ subshell
of the third shell and thus is
represented as 3s1

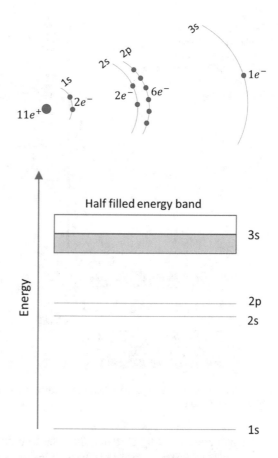

Fig. 12.4 As there is only
one valence electron, the
'3s' energy band in solid
sodium is therefore only half
filled

energy band. If N atoms have joined together to form the $3\,s$ band, then it can hold a
total of $2\,N$ electrons. As only N electrons reside in the $3\,s$ band, it is therefore only
half filled by electrons. The inner levels, viz., $2p$, $2\,s$, and $1\,s$, being closer to the
nucleus, show insignificant broadening of the energy levels. This situation has been
described in illustration of Fig. 12.4. Even when a modest potential difference is
applied across a block of solid sodium, the $3\,s$ band electrons can readily acquire
energy while still remaining in the same band. The ensuing drift of these valence
electrons therefore constitutes a net electric current. It is no wonder that sodium is a
good conductor of electricity and is accordingly classified as a metal.

In the case of an insulator, on the other hand, the outermost subshell is completely
full, and in turn, the valence energy band here is thus fully occupied (Fig. 12.5). The
next outer subshell that is completely empty, and has the potential to function as a
conduction band, is separated from this valence band by a forbidden energy gap
often measuring several electron volts.[4] The average thermal energy under ambient
conditions being only about 0.025 eV, the valence electrons are unable to appear in

---

[4] 1 eV of energy equals 1.6×10^{-19} joule.

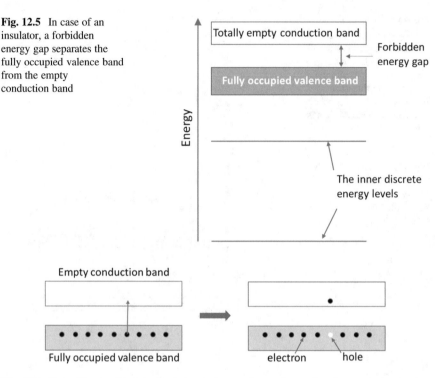

Fig. 12.5 In case of an insulator, a forbidden energy gap separates the fully occupied valence band from the empty conduction band

Fig. 12.6 The void created in the valence band when an electron jumps from here to the conduction band is called hole

the conduction band by climbing over the forbidden energy barrier. Nor can they cross such an energy barrier by the application of an electric field of reasonable magnitude that readily sets in a current to flow in a metal. As a matter of fact, an electric field of magnitude to the tune of tens of million V/m is needed for an electron to jump from the valence to conduction band in a typical insulator.

There is yet another kind of substance called a semiconductor that also has a fully occupied valence band like an insulator, but quite unlike it, the empty conduction band here is located only marginally above the valence band; the forbidden energy gap here can be ~1 eV or even less. Such a material behaves like an insulator at low temperature, but at room temperature or above, a small fraction of electrons in the valence band has enough thermal energy to cross over the narrow energy barrier and enter the conduction band. Although small in number, these electrons are still good enough to manifest as a flow of current in the conduction band upon application of a reasonable electric field. The story doesn't end here though, as in contrast to a metal, the semiconductor carries current not only in the conduction band but also in the valence band. The electron, by jumping from the valence to the conduction band, leaves behind an unoccupied state in the valence band called a *hole* (Fig. 12.6). These holes can basically be regarded as a deficit of electrons or equivalently as a positive charge. Just as the motion of electrons in the conduction band results in the

Fig. 12.7 With passage of time as the electrons move from left to right, the hole moves from right to left

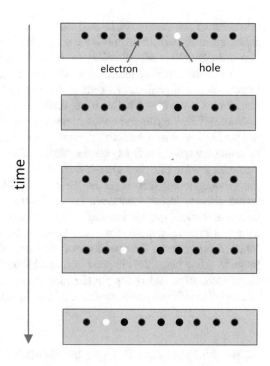

flow of a current, the flow of current in the valence band too arises from the motion of holes here. In reality, however, the holes do not move by themselves but appear to be doing so as the valence band electrons basically engage themselves in a game of musical chairs. This point can be readily understood by referring to Fig. 12.7, which essentially represents a collection of snapshots of the motion of the electrons in the valence band captured sequentially in time. Following the creation of a hole in the valence band, one of the neighboring electrons, decided by the polarity of the applied field, moves quickly into it, thereby creating a hole in its original location. As a second electron now moves to fill this new hole, it leaves behind yet again a hole and so on. In a nutshell, therefore, motion of the hole occurs in a direction opposite to that of electrons. Thus, in a semiconductor, the flow of current in the valence band can as well be thought of as originating from the motion of the holes instead of the electrons. This hole-centric approach is also inherently simple to analyze as in the valence band there are fewer holes that are greatly outnumbered by the electrons.

12.3 Impurity Semiconductors

Semiconductor materials exist both in elemental and compound forms. The two most popular element-based semiconductors are silicon and germanium. Some of the molecular variety of semiconductors that find wide applications are GaAs, GaP, InP, etc. (GaAs and InP, as a matter of fact, are the most widely used materials for the operation of semiconductor lasers.) Both these varieties of semiconductors are also called intrinsic semiconductor as the semiconducting properties are inherent to these materials. By virtue of their ability to conduct electricity by both electrons and holes, semiconductors display remarkable properties; viz., the presence of only a trace of impurities can significantly enhance the conductivity of a semiconductor. Semiconductors embedded with such impurities are called *doped* or *extrinsic* semiconductors. While varying the concentration of the dopant in a semiconductor allows ready tailoring of its conductivity, careful selection of the dopant material will allow selecting the type of majority charge carriers. The extrinsic semiconductor for which electrons are the principal charge carriers is called an n-type semiconductor, while the semiconductor with holes as the majority charge carriers is termed a p-type semiconductor. A semiconductor reveals its true potential when a p-type semiconductor and an n-type semiconductor are brought in contact with each other. Interesting effects originating from the passage of the charge carriers through this junction bring about the operation of a variety of semiconductor devices, the p-n junction diode being the simplest of them. In the following section, we shall take a closer look at the underlying physics that allows operation of a semiconductor diode as an emitter of both conventional and laser lights.

12.4   N-Type and P-Type Semiconductors

To understand the construction and operation of these type of semiconductors, let us begin with crystalline silicon as the fundamental intrinsic semiconductor material. A silicon atom has four electrons in its outermost shell, and they, as shown in Fig. 12.8,

Fig. 12.8 In a crystalline silicon, there is no loose or free electron as all the outer shell electrons are engaged in forming bond with the neighbors

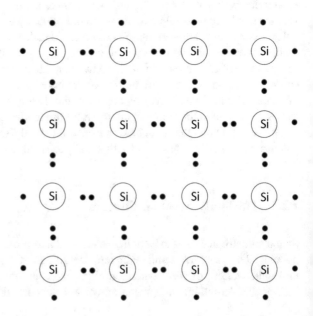

Fig. 12.9 Upon doping, as an arsenic atom displaces a silicon atom in the crystal, the fifth outer shell electron of As with no bonding is now loosely bound. The ability of these electrons to move freely in the crystal dramatically enhances its conductivity

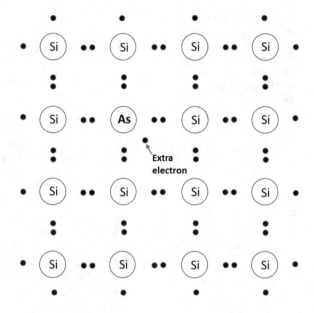

Fig. 12.10 Band structure of an n-type semiconductor revealing clearly that electrons are the majority charge carrier here

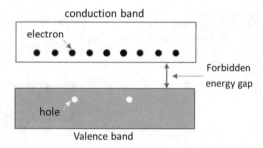

are all individually bonded to the electrons of the adjacent atoms while forming the crystal. In the absence of any loose or free electron, the silicon crystal will not conduct electricity unless some electrons from the valence band are excited to the conduction band either thermally or by some other means. However, doping of this silicon crystal with trace quantities of arsenic in a ratio of about 1 in 100 million brings about remarkable changes in its electrical conductivity for the following reason. As a result of doping, the arsenic atoms will replace silicon atoms at a few locations in the crystal. An arsenic atom has five electrons in its outermost shell of which four will be engaged in forming bonds with their nearest *Si* neighbors. The remaining electrons being loosely bound can therefore move freely in the crystal as shown in Fig. 12.9. The arsenic atoms function here as the *donor* of electrons that can easily get into the conduction band. This situation has been schematically illustrated in Fig. 12.10. As seen, the majority of carriers in this extrinsic semiconductor are electrons and hence the name *n*-type semiconductor. The few holes that are present in the valence band owe their origin to the direct excitation of a few

Fig. 12.11 When the silicon crystal is doped with gallium, deficit of electron results in the creation of hole that can move freely in the crystal boosting its conductivity

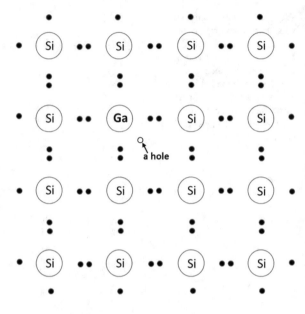

Fig. 12.12 Band structure of a p-type semiconductor revealing clearly that holes are the majority charge carrier here

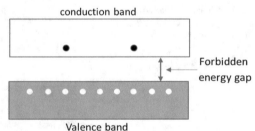

electrons from the valence to the conduction band. Needless to say, most of the electrons in the conduction band have been donated by the arsenic atoms.

Doping the silicon crystal with trace quantities of gallium atoms, on the other hand, results in a totally different behavior. A gallium atom has only three electrons in the outer shell, and as before, they will replace silicon atoms at a few locations in the crystal. It is not difficult to perceive that as a gallium atom gets electronically bonded with the neighboring silicon atoms, a hole, as depicted in Fig. 12.11, will be formed where the fourth electron should have been present. As already explained, these holes will conduct electricity in the valence band, and the situation has been described in Fig. 12.12. In contrast to the previous case, the majority charge carriers are now holes, and accordingly, this variety of extrinsic semiconductors is termed *p*-type semiconductors. The few electrons present in the conduction band originate from the thermal excitation of some electrons from the valence to the conduction band that, as we know, occurs in an intrinsic semiconductor at room temperature. The gallium atoms thus function here as acceptors of electrons making the movement of the holes possible in the valence band.

12.5   Semiconductor Diodes

Joining together the p and n junction semiconductors in the manner shown in Fig. 10–38 added a new dimension to the semiconductors paving the way for their device applications. It is worthwhile at this point to take a deeper look at the working of a p-n junction. Left to itself, some of the free electrons from the n-region will diffuse toward the p-region, and likewise, holes from the p-side of the junction will diffuse toward the n-side (Fig. 12.13a). Electrons exist in the conduction band, while the holes are in the valence band, and as they meet at the junction, the electrons jump into the holes, and both carriers disappear. This leaves in the immediate vicinity of the junction a layer of positive ions in the n-side and a layer of negative ions in the p-side, as shown in Fig. 12.13b. The charge separation (between the immovable positive and negative ions) sets up a field that opposes any further diffusion of the charge carriers across the junction. Soon, an equilibrium will be established when the opposing field will be strong enough to put a stop on any more diffusion. This results in establishing a thin region in the vicinity of the junction that is devoid of any carriers, called a depleted region, opening up the interesting possibility of using the p-n junction as a diode. To understand this, we need to consider two cases. First, we connect the n-side to the positive terminal of a battery and the p-side to its negative terminal, as shown in Fig. 12.14a. There will be a momentary flow of current as the free electrons from the n-region and the free holes from the p-region flow away into the battery. This quickly results in the depletion of charge carriers on either side of the junction, thereby causing stoppage of the flow of current through the circuit almost instantaneously. However, the presence of some residual current, which is the reverse current, cannot be ruled out arising from the creation of occasional electron-hole pairs owing to thermal fluctuation that is inevitable. This thus represents operation of the p-n junction under reverse bias condition. Next, we reverse the polarity of the battery and the operation here is shown in Fig. 12.14b. The battery now clearly pumps electrons into the n-region and removes them from the p-region,

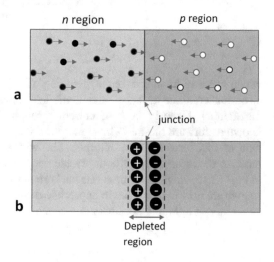

Fig. 12.13 When an n-type semiconductor is joined with the p-type, the loosely bound electrons move from the n to p region. The loosely bound holes, on the other hand, move in the reverse direction (**a**). As the holes and electrons meet at the junction and annihilate each other, a layer of positive and negative ions is formed, respectively, in the n- and p-sides of the junction in the equilibrium. This results in the formation of a thin layer of charge depletion region around the junction (**b**)

Fig. 12.14 The reverse
(**a**) and forward (**b**) bias
operation of a p-n junction

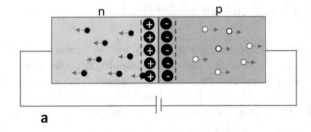

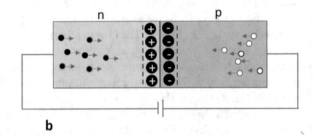

Fig. 12.15 The current as a
function of voltage in the
operation of a p-n junction
diode. The small residual
current in case of reverse
biasing of the diode is
clearly evident

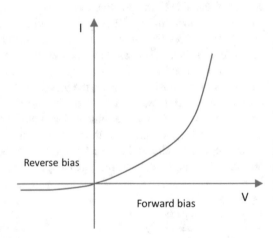

which can also be interpreted as pumping of holes into this region. The holes and
electrons continue to meet and annihilate each other in the junction region,
maintaining a steady flow of current through the circuit in this configuration
known as forward biasing. The higher the battery voltage is, the greater the magnitude of this forward biased current. The voltage-current characteristics of a p-n
junction diode operated in both reverse and forward biased conditions are shown
in Fig. 12.15. Having acquired this background knowledge of semiconductor physics, we are now in a position to appreciate the working of a semiconductor diode as
an emitter of light, both ordinary and coherent.

12.6   Light Emitting Diodes (LED)

Let us in the beginning take a closer look at the creation and annihilation of an electron-hole pair in a semiconductor. We know that a minimum amount of energy equaling the bandgap energy is required to be imparted to a valence band electron for its excitation to the conduction band leaving behind a hole in its original location. Although the omnipresent thermal energy normally facilitates this excitation, it can also be achieved optically provided the photon energy exceeds the bandgap energy. When we say that an electron and hole combine to annihilate each other, what basically happens is that an electron from the conduction band falls into a hole at the valence band and the two nullify each other. As shown in Fig. 12.16, the excess energy that is released when the electron jumps down into the valence band can be emitted as a photon.

Considering the operation of the *p-n* junction, we know that upon forward biasing, it allows a flow of current, while the holes and electrons continue to recombine at the junction and nullify each other. Now recombination is nothing but the conduction band electron literally jumping into the hole in the valence band. If the excess energy of the electron is released as a photon, the diode can then be said to be a light emitter. We can thus say that a *p-n* junction, when forward biased, is capable of emitting light. There is however a catch here. The electron-hole transition is radiative normally for only direct bandgap semiconductors. In a direct bandgap semiconductor, the total momentum of the electron and hole do not change as they

Fig. 12.16 Creation of a photon as an electron and a hole meet and annihilate each other. The electron and the hole originally reside, respectively, in the conduction and valence bands (**a**). As the electron jumps down into the hole (**b**), both disappear and the excess energy carried by the electron is emitted as a photon (**c**)

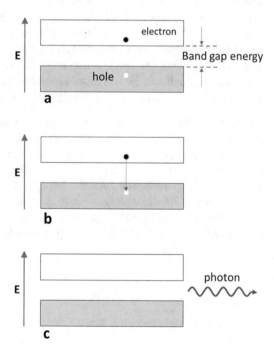

recombine. The transition only involves a change in energy that is carried by the emitted photon. In an indirect bandgap semiconductor, on the other hand, the total momentum of the electron and hole changes as they recombine. Conservation of momentum is not possible if the excess energy is released only as a photon. The electron needs to pass through an intermediate state to transfer momentum to the crystal lattice, and in turn, the energy is dissipated as heat instead of light. Silicon, the best-known semiconductor material that serves as the building block of miniature electronics, is unfortunately an indirect bandgap semiconductor and thus does not make a good light emitter and hence an LED. Other semiconductors, such as germanium, GaP, AlP, AlAs, and AlSb, also have indirect bandgaps and thus perform poorly as emitters of light. Some of the notable examples of a direct bandgap semiconductor are GaAs, GaSb, InAs, InP, and InSb, of which GaAs has the distinction of emerging as the very first light emitting diode.

12.7 Diode Lasers

Thus, there exists a semiconductor device that is capable of emitting light. The emission here is spontaneous in nature, and the emitted light is therefore ordinary and spreads out in all possible directions. It is not difficult to figure out that a light emitting diode can be converted into a laser by realizing in it the condition of population inversion and simultaneously ensuring the presence of optical feedback. The population inversion can be ensured by appropriately increasing the pumping current so that the rate at which electrons fall into holes and recombine exceeds the rate at which electrons at the valence band can jump into the conduction band by absorbing photons emitted from the recombination. Feedback, as we know, can be provided by placing the device inside an optical cavity. Nature has been kind here, since as the refractive index of semiconductor materials is usually high (e.g., $\mu = 3.6$ for GaAs), the Fresnel reflectivity[5] $\left[\frac{(\mu-1)^2}{(\mu+1)^2}\right]$ is therefore also quite high (~32% for GaAs). Cleaving the end faces perpendicular to the laser axis will therefore allow much of the light created in the junction to be reflected back and forth. This will help meet the threshold condition for lasing in many cases. The configuration of a typical diode laser is schematically illustrated in Fig. 12.17. As it occurs in any other laser, a fraction of the spontaneously emitted photons in the junction that travel along the axis will initiate the stimulated emission causing amplification of light. For low gain lasers or when a unidirectional laser beam is essential, the reflectivity of the faces can be manipulated by appropriately coating them with dielectric layers.

---

[5] Readers may take another look at Sect. 2.3.1 of Chap. 2 for refreshing their memory on Fresnel reflection.

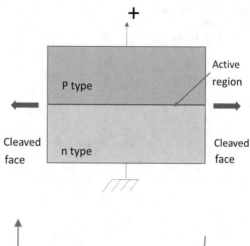

Fig. 12.17 A light emitting diode turns into a laser when the forward biased current exceeds the lasing threshold value. This is a longitudinal or side view of the laser

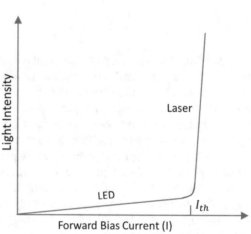

Fig. 12.18 Switching of the operation of a forward biased semiconductor diode from LED into laser with increasing bias current. 'Ith' is the threshold value of current beyond which the diode works as a laser

12.8 Homojunction Diode Lasers

When the diode is made out of the same type of semiconductor material on both sides of the junction, the laser is called a homojunction laser. When the diode is forward biased and the current begins to flow, recombination of electron-hole pairs sets in across the junction resulting in the emergence of spontaneously emitted light. As schematically shown in Fig. 12.18, the intensity of this light gradually increases with the current. This monotonous rise in the intensity of light continues until the current reaches a critical value when it exhibits a dramatic rise, revealing the onset of lasing. The value of the current at which the amplification of light due to stimulated emission begins is called the threshold current for lasing. The requirement of this threshold current is exorbitantly high for a homojunction semiconductor laser for the following reason. Let us consider a GaAs diode, and an appropriately labeled cross-sectional view of the same is shown in Fig. 12.19. Two sides of a GaAs block are doped, the top face as p and the bottom face as n. When the diode is forward biased,

Fig. 12.19 Cross-sectional
or end view of a
homojunction diode

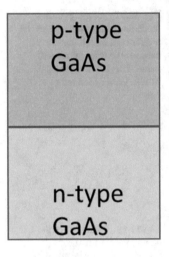

electrons from the *n*-region and holes from the *p*-region are injected into the junction where they recombine and emit light. As it is GaAs everywhere, this light can easily escape through both the top and bottom faces of the junction. Furthermore, there is nothing to confine the holes and electrons into the junction to ensure a better rate of recombination. These two factors together contribute to the extremely low electro-optic efficiency of a homojunction diode laser. This results in dissipation of excessive heat that can easily damage the diode. To prevent this, it is mandatory to cool the device by placing it in a liquid nitrogen bath (~77° K). However, even with such cooling, they are capable of operating only in the pulsed mode as *cw* operation is impossible to achieve. Although historically significant and easy to understand, these homojunction lasers, with their extra baggage of liquid N_2 bath, are unsuitable for any practical applications.

12.9 Heterojunction Diode Lasers

The limitations of the homojunction lasers can be overcome if both the light and current carriers are confined within the narrow zone of the junction layer, a seemingly uphill task. It is no wonder that the operation of a *cw* diode laser was realized at room temperature more than half a decade after the discovery of lasing in a homojunction diode in the latter half of 1962. Nature was again kind here as a material with a higher bandgap intrinsically possesses a lower refractive index.[6] This

---

[6]In 1952, Trevor Simpson Moss (1921–), a British physicist, published a book (T.S. Moss, Photoconductivity in the Elements, Academic Press Inc., New York, 1952) during his affiliation to Cambridge University, connecting the electronic and optical properties of a semiconductor by a rule that is known as Moss rule and is given by *Band Gap energy × refractive index = constant*.

Fig. 12.20 A semiconductor material of higher bandgap has a smaller r.i. and vice versa. A heterojunction therefore neither allows passage of current nor light (when angle of incidence exceeds critical angle) from the lower to higher bandgap side

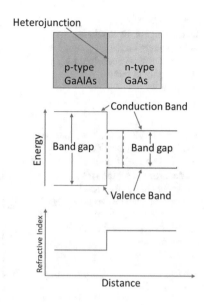

advantage was exploited in 1968 by Herbert Kroemer (b–1928), a German-American physicist, and Zhores Alferov (1930–2019), a Soviet and Russian physicist, when they succeeded independently in the confinement of both light and current inside the junction layer. The conception of a novel heterojunction diode did the trick here, a masterstroke that killed two flies with one slap. The room temperature operation of the *cw* diode laser eventually fetched Kroemer and Alferov the 2000 Nobel Prize in Physics.

In order to simplify the understanding of the physics of confinement, we consider at this point a single heterojunction diode as shown in the top trace of Fig. 12.20; *p*-type GaAlAs and *n*-type GaAs make a heterojunction with it. GaAlAs is a higher bandgap semiconductor, and the electrons that are injected into the junction from the *n*-side lack energy to cross over into it as GaAs is a lower bandgap semiconductor. This situation is depicted in the middle trace of this figure. This, in turn, results in an improved confinement of the charge carriers in a single heterojunction laser. It readily follows from Moss's rule that GaAs, being a lower bandgap material, will have a refractive index exceeding that of GaAlAs. Consequently, the passage of light from GaAs to GaAlAs will be blocked as the phenomenon of total internal reflection will be operative here. A single heterojunction laser thus allows only partial confinement of current and light and is capable of operating in the pulsed mode at room temperature. The confinement would be much better by going for a double structure heterojunction diode, as shown in Fig. 12.21. As seen, a thin slice of intrinsic GaAs semiconductor is basically sandwiched here by *p*-type GaAlAs on the left and *n*-type GaAlAs on the right. The double-heterojunction diode laser with improved confinement of both current and light within the lasing volume therefore ensures that the majority of the charge carriers recombine here and emit light. As shown in Fig. 12.22, the threshold current for the onset of lasing is therefore remarkably less

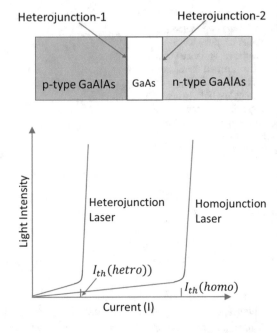

Fig. 12.21 A double-heterojunction diode when appropriately biased is endowed with the capability of cw lasing at room temperature

Fig. 12.22 The better confinement of current and light in a heterojunction diode laser lowers the threshold current for lasing significantly. The behavior of a homojunction laser is also shown here for the sake of comparison

than that for the homojunction lasers. This results in a significant improvement in the performance of the laser by establishing increased electro-optic efficiency and reduced heat dissipation. It is not surprising that room temperature *cw* operation is readily achievable from a double-heterojunction laser. As a matter of fact, the semiconductor lasers of today are primarily based on such double-heterostructure design of the diodes.

12.10   Quantum Well Lasers

Double-heterojunction lasers, we now know, offer lower lasing threshold current and higher electro-optic efficiency primarily because they confine the charge carriers in the active layer between the junctions. This enhances the probability of electrons and holes recombining and emitting light that undergoes total internal reflection within this double heterostructure and cannot readily escape. Quite naturally, therefore, the thinner the active layer is, the more likely the recombination, and the better the laser performs. With the advancement of layer deposition technology, it is now possible to fabricate semiconductor layers as thin as tens of nanometers. At such a small thickness that is comparable to the de Broglie wavelength of the electron, the quantum size effect sets in and profoundly impacts the operation of the diode lasers. The electrons now behave like a particle in a quantum well, resulting in quantization of the energy levels as shown in Fig. 12.23. This is akin to the case of a particle in a box leading to energy quantization and has been elaborated in Chap. 3. Exactly the

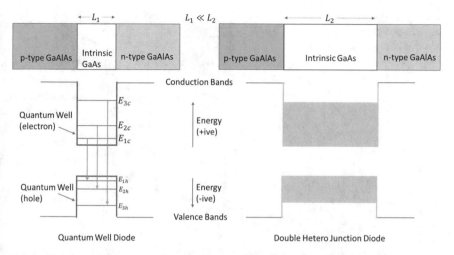

Fig. 12.23 When the thickness of the junction layer approaches the de Broglie wavelength of the electron, the energy levels are quantized for both electrons in the conduction band and holes in the valence band. The conduction and valence energy bands in case of a double heterojunction are also shown here. The thickness of the active layer has been exaggerated to clearly distinguish the two diodes

same happens to the energy levels of the holes in the valence band as well but only in a direction opposite to that in the conduction band forming an upside down well. Holes being heavier than the electrons, energy levels are more closely packed in the case of holes. Only a few low-lying energy levels usually lie inside the quantum wells, and the ones that lie outside bear no consequence in the context of quantum well lasers as the electron or hole with such energies will essentially escape the well. We know from Eq. 3.4 , $E_n \propto \frac{n^2}{mL^2}$, where n, L, and m are the level quantum number, thickness of the active layer, and mass of the charge carrier, respectively. The quantized values of energy can therefore be appropriately tailored to suit the requirement of a particular application by adjusting the layer thickness. The allowed transitions are those for which the n value does not change, and the same are also shown in this figure. With reducing L, the energy levels move away from each other, while with increasing L, the levels approach each other. When L far exceeds the de Broglie wavelength of the electron, the levels become so closely packed as to turn into an energy band, the signature of the double-heterojunction diode, also shown on the right side of this figure for the sake of comparison.

The operation of a quantum well laser is illustrated in Fig. 12.24. A slice of GaAs intrinsic semiconductor sandwiched between a p-type and an n-type GaAlAs semi-conductors forms the active layer of the quantum well diode considered here. The extreme narrowness of the active layer allows only a lone energy level to reside inside both the conduction and valence band quantum wells. When the diode is forward biased, electrons and holes injected into the n- and p-regions of the diode by the battery appear in the energy levels of the respective quantum wells. The electrons and holes can now recombine to produce light with remarkable ease in the ultrathin

Fig. 12.24 Schematic representation of the working of a quantum well laser

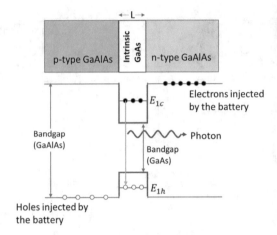

active layer. The inherent high electro-optic efficiency of a quantum well laser allows realization of population inversion and, in turn, lasing at a modest value of the bias current.

Charles H. Henry (1937–2016), an American physicist, conceived the idea of discretizing the energy levels of semiconducting electrons by confining them inside a quantum well in late 1972, while he was heading the Semiconductor Electronics Research Department of Bell Labs, Murray Hill, New Jersey. Later, in 1974, in association with Raymond Dingle, a physicist from the same department, he operated the first quantum well laser[7] and promptly secured a US patent entitled "Quantum Effects in Heterostructure Lasers." The very thin heterostructure, the heart of this device, was fabricated by W. Wiegmann using the technique of molecular-beam epitaxy.[8] Although invented at Bell Laboratories, the name "quantum well laser" was coined by Nick Holonyak, the inventor of LED, and coworkers at the University of Illinois, Urbana-Champaign.

A quantum well laser offers a number of distinct advantages. The first and the foremost is the ease with which the lasing wavelength can be tuned by changing the thickness of the active layer. This can be understood by taking another look at Fig. 12.23. The wavelength of the photon emitted from the radiative transition between energy levels E_{1c} and E_{1h} is essentially governed by the energy difference between these two levels. As the energy quantization and the thickness of the layer are interconnected, the wavelength of the emitted photon will also depend on the layer thickness. More precisely, the wavelength decreases with the reduction in layer thickness and vice versa. To effect such a change in thickness, even on a minute

---

[7]To know the story of the invention of the quantum well lasers in the inventor's own word, the readers may refer to the foreword that C. H. Henry wrote for the book *Quantum Well Lasers* ed. by Peter S. Zory Jr.

[8]Molecular-beam epitaxy (MBE) is a process for thin film deposition of single crystals. The ultrahigh vacuum and extremely low deposition rate adopted here allow exquisite control over material purity and layer thickness, respectively.

Fig. 12.25 Schematic representation of stacking four quantum wells together to fabricate a multi QW laser

scale, is quite straightforward in practice, e.g., in the MBE, which is widely used in the manufacturing industry of semiconductor devices, the layer thickness can be precisely maneuvered by controlling the thin film deposition time. In contrast, wavelength tuning, albeit discretely, in a typical diode laser is possible only by changing the energy bandgap that mandatorily calls for changing the layer composition, a more complex and involved procedure nevertheless.

As in the greatly reduced volume of the active layer electrons and holes can find each other and recombine with remarkable ease, a quantum well laser thus requires fewer holes and electrons to reach the lasing threshold compared to double-heterojunction lasers. This is the primary reason as to why a QW laser can perform at a high electro-optic efficiency even though the reduced active volume greatly lowers the mode filling factor.[9] The small active volume, however, limits the maximum optical power that can be extracted from a single QW laser. This disadvantage can be overcome by stacking in parallel a few identical quantum well heterojunction structures with a separation judiciously chosen so that they are optically coupled but electronically noninteracting. The nonoverlapping electron wave functions ensure that the energy quantization of individual heterostructures remains unaffected giving rise to multiple quantum well structures, as shown in Fig. 12.25. The strong optical coupling between the individual active layers in conjunction to their integration within a single optical cavity ensures, on the other hand, an output beam of good spatial quality. In summary, the laser now generates an output beam of higher power in accordance with the increased active volume as a result of stacking multiple quantum well structures but with seemingly no ill effect on its spatial quality.

12.11 Quantum Cascade Lasers

Unlike typical semiconductor lasers wherein light originates from inter-band transitions involving the recombination of electrons and holes, a quantum cascade laser operates on intra-band sublevels involving only one type of carrier, viz., an electron. To drive this point home, the emission of a QC laser has been illustrated in Fig. 12.26 alongside that of a QW laser. Although the possibility of a quantum cascade laser

---

[9]In the operation of a laser, the oscillating mode usually does not have a complete overlap with the active volume of the lasing medium. The fraction of the gain volume that the mode utilizes for its growth is called the "mode filling factor."

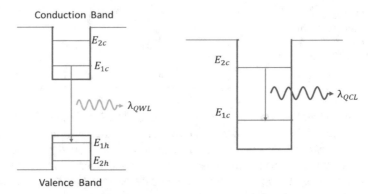

Fig. 12.26 Quantum cascade laser is purely of intra-band origin where the electrons make a transition between the discrete energy levels within the conduction band quantum well (right trace). In case of a quantum well laser, on the other hand, the emission originates from the electron-hole recombination across the material bandgap (left trace)

was conceptualized in 1971, it took more than two decades before such a laser was operated for the first time at the Bell Laboratories, Murray Hill, New Jersey, by Alfred Y. Cho (1937–) and Federico Capasso (1949) and coworkers. As occurs in the case of QW lasers, here too an ultrathin active layer leads to quantum confinement of the electrons resulting in the quantization of their energy levels. The spacing between these levels, as we know, depends on the width of the active layer and can thus be altered by varying the same. The electrons can jump from a higher energy level to a lower energy level releasing the excess energy as a photon. The QC lasers are capable of emitting light of much longer wavelength than achievable for a QW laser, where the lasing transition occurs between widely separated energy levels as one lies in the conduction band QW and the other lies in the valence band QW. These lasers are therefore capable of emitting on the mid- and far-infrared regions of the e-m spectra with wavelengths ranging from several micrometers to hundreds of micrometers and well into the THz region.

As the name suggests, the operation of a QC laser is realized by laying a series of quantum wells that are appropriately biased to essentially form an energy staircase. As the electron tunnels[10] resonantly from one quantum well to the next and travels down the energy staircase, much like a marble rolling down the stair, it emits a photon every time it cascades down an energy step. The situation is schematically described in the illustration of Fig. 12.27. As seen, a single electron, upon its injection into the gain region, is made to emit multiple photons. The number of wells to be coupled is essentially governed by the requirement of the optical power and can vary from a few to tens to even a hundred depending on the applications.

---

[10]The quantum tunneling effect is the ability of a particle to penetrate a potential barrier with an energy less than the height of the barrier. Exponential decay of the magnitude of the wave function of the particle allows its quantum mechanical penetration through the barrier.

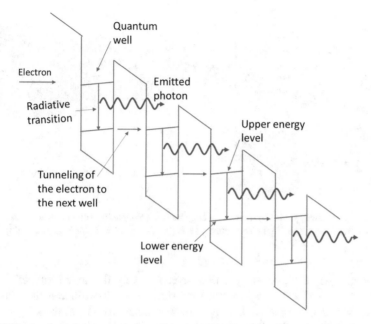

Fig. 12.27 A schematic representation of the principle of working of a QC laser. A judicious combination of the layer thickness and the applied electric field makes possible for the electron to resonantly tunnel to the next well. Consequently, a single electron is able to manufacture multiple photons as it cascades down the energy steps

The inbuilt capability of readily tunable long wavelength operation has made these lasers particularly suitable for trace gas analysis and chemical sensing applications, and the same will be elaborated in volume II of this book.

12.12   Edge and Surface Emitting Diode Lasers

So long our focus has been largely on the variety of compositions and configurations of the heterojunction diode and its active layer with no special emphasis or consideration on the arrangement of the optical cavity. We now turn our attention to the organizations of the optical cavity to determine their bearing on the overall performance of the heterojunction lasers. The early work relied on the cavity with longitudinal geometry that offered two inherent advantages: (1) The active length here equals the length of the junction layer, and (2) the two cleaved end facets with intrinsically high Fresnel reflectivity can perform the role of the cavity mirrors. A laser based on such a cavity would thus emit through its ends and is therefore termed as an edge emitting laser. A typical edge emitting laser with GaAs as the base semiconductor material has been schematically illustrated in Fig. 12.28. As seen, the length and width of the n-GaAs substrate are approximately 300 and 200 μm,

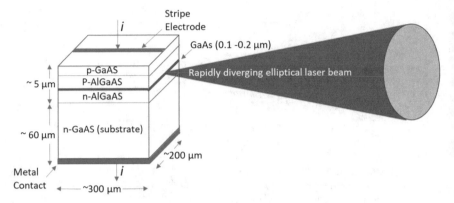

Fig. 12.28 The schematic illustration of an edge emitting double heterojunction diode laser. The ultrathin active layer results in the generation of a highly diverging beam of light that is elliptical in shape

respectively, while the GaAs junction layer is barely 0.1 to 0.2 μm thick. The *p*-AlGaAs and *n*-AlGaAs layers lying on either side of GaAs constitute the *p-n* junction. In order to establish a proper ohmic contact, the ~1-μm-thick top layer is made out of *p*-type GaAs. To restrict the current flow through a narrow region to increase the density of charge carriers, the electrical connection is taken by laying a stripe electrode atop. AlGaAs being a higher bandgap material than GaAs, there will be, as we know, an automatic confinement of both the charge carriers and light within the active volume. Having gone through the phenomenon of the diffraction of light in Chap. 2, we now know that the divergence of a beam of light is primarily governed by its size at the source – the narrower the source, the higher its divergence and vice versa. The exceptionally thin active layer of the diode laser (that usually lies within 0.1 to 0.2 μm and is considerably lower than the wavelength of emission) therefore makes the beam emerging through it extremely divergent. As the active zone is smaller in the vertical direction than in the horizontal direction, the divergent beam is thus essentially elliptical in shape. Although the beam can be corrected externally to some extent, this strong diverging component intrinsic to the edge emitting diode laser nevertheless poses a major challenge.

This limitation of edge emitting lasers can be readily overcome by configuring the cavity in the transverse direction, as it allows the ready advantage of greatly enlarging the emitting area of the laser. Popular as VCSEL (vertical cavity surface emitting laser), a beam with remarkably improved spatial quality, which is both round and much less divergent, emerges from such a laser. The arrangement of a VCSEL has been conceptually depicted in the illustration of Fig. 12.29. As can be readily appreciated, this advantage comes at the expense of the gain length, which is now barely a fraction of a micrometer equaling the thickness of the active layer. This greatly reduced active length can be offset by accordingly reducing the round-trip cavity loss. The moderately high reflectivity that the semiconductor materials are known to possess and comes in handy for edge emitting lasers would not thus suffice

Fig. 12.29 Conceptual depiction of a surface emitting double heterojunction diode laser. Participation of a significantly larger and circular cross section of the active layer into the process of lasing results in the emission of a less divergent and round output beam

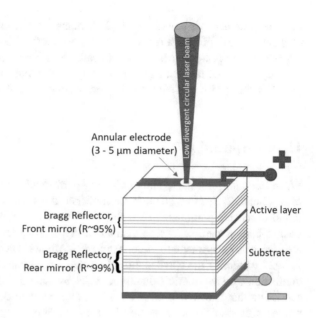

here. The use of Bragg reflectors,[11] which can offer remarkably high reflectivity when appropriately designed, has emerged as the most attractive choice to effect lasing in these vertical cavity lasers. As seen, the illustration of Fig. 12.29 makes use of Bragg mirrors of reflectivity ~99% and ~ 95% as rear and front mirrors, respectively, to form the resonator cavity. The top electrode is made annular to allow emergence of the low divergent circular output beam. VCSELs are inherently low-power devices and usually generate less power than edge emitting lasers as the active volume where recombination of charge particles takes place is much too small. These lasers, with their very low lasing current threshold and satisfactory beam quality, are therefore an attractive choice for low-power applications.

12.13   Laser Diode Arrays: From Watts to Kilowatts

The tiny physical size of the active layer in a diode laser essentially limits its power generation capability to a modest level, which may be adequate for a host of the aforementioned applications. However, for applications requiring higher power such as pumping high-power lasers or a variety of material processing and medical applications, power scaling is possible by combining the output of many single emitter devices. The number of diodes to be added depends on the requirement of power, and they can be stacked in a linear array often called a diode bar. To meet the

---

[11] A Bragg reflector essentially is a mirror structure which consists of a stack of alternating high and low index layers of two different optical materials.

increasing demand of power, the diode bars can be stacked side by side as well as atop each other, forming a regular matrix of emitters with the capability of generating continuous power exceeding a kilowatt. Needless to say, provision of adequate cooling by way of fabricating microchannels within the entire structure is a prerequisite for the working of such a high-power device called a diode array.

12.14   Appendix

Wave Function: In 1801, Thomas Young performed the famed double slit experiment (Chap. 2, Section 2.4) that reestablished the wave nature of light. Light was found to reach a point behind the slits that would be out of bounds if light were to behave here as particles. More than a century and quarter later, the interference of electron was experimentally demonstrated by Clinton Davisson and Lester Germer in 1927 validating the wave nature of microscopic particles, soon after the conceptualization of wave-particle duality in 1924 by Luis de Broglie. It is therefore not surprising that Newtonian mechanics is not adequate to describe the dynamics of a microscopic particle, such as an electron, and this is where quantum mechanics comes in handy. The electron behavior, assuming it to be travelling in a free space, can now be satisfactorily described by Schrodinger's wave equation

$$i\hbar\frac{\partial \Psi(x,t)}{\partial t} = -\frac{\hbar^2}{2m}\frac{\partial^2 \psi(x,t)}{\partial x^2}, \text{the quantum equivalent of Newton's law of motion} \quad (12.1)$$

where $\hbar$ is the reduced Planck's constant and m is the mass of electron. More importantly, the function $\psi(x,t)$ is identified as the electron's wave function which is complex valued for every position x and for all times t. The wave function of a particle at a given position and time, therefore, only represents the likelihood of finding the particle at that location and time and has no physical significance. By analogy with, say, an acoustic wave, it can be thought to be an expression for the probability amplitude of the de Broglie wave of the electron. The absolute value squared of the wave function ψ, viz., $|\psi|^2$, being real, however, has a physical significance. It represents the probability density of finding a particle in a given region of space and at a given point in time.

References

1. P. Lenard, Ueber die lichtelektrische Wirkung. Ann. Physik **8**, 149 (1902) [This is in German, Interested readers can refer to 'B. R. Wheaton, "Philipp Lenard and the Photoelectric Effect, 1889–1911", Historical Studies in Physical Sciences, Vol – 9 (1978)' for a more comprehensive account on the discovery of photoelectric effect]
2. J.D. Norton, Chasing a Beam of Light: Einstein's Most Famous Thought Experiment, https://www.pitt.edu/~jdnorton/Goodies/Chasing_the_light/
3. A.A. Mills, R. Clift, Reflections of the 'Burning mirrors of Archimedes'. With a consideration of the geometry and intensity of sunlight reflected from plane mirrors. Eur. J. Phys. **13**, 268 (1992)
4. N. Bohr, On the constitution of atoms and molecules. Philos. Magaz. Ser. 6 **26**(151), 1–25 (1913)
5. A. Einstein, On the quantum theory of radiation. Phys. Z. **18**, 121 (1917)
6. M. Rose, H. Hogana, History of the Laser: 1960–2019. https://www.photonics.com/Articles/A_History_of_the_Laser_1960_-_2019/a42279
7. J. Hecht, *Laser Pioneers*, ISBN 0 12-336030-7 (Academic, 1991)
8. T.H. Maiman, Stimulated optical radiation in ruby. Nature **187**, 493 (1960)
9. J. M. Vaughan (1989). The Fabry-Perot interferometer: history, theory, practice, and applications, New York, eBook (2017), ISBN 9780203736715
10. Laser Odyssey, Theodore Maiman, ISBN 0-97-029270-8, Laser Press, Blaine, WA (2000)
11. https://www.laserfocusworld.com/test-measurement/research/article/16565577/smallest-semiconductor-laser-created-by-university-of-texas-scientists
12. https://www.forbes.com/sites/meriameberboucha/2017/10/29/worlds-largest-laser-could-solve-our-energy-problems/?sh=444cfe754ffd
13. https://www.kent.edu/physics/top-10-beautiful-physics-experiments
14. A.M. Sayili, The Aristotelian Explanation of the Rainbow. Isis **30**(1), 65–83 (1939) (19 pages), Published By: The University of Chicago Press. https://www.jstor.org/stable/225582
15. https://www.forbes.com/sites/startswithabang/2018/08/16/the-greatest-mistake-in-the-history-of-physics/?sh=518c7a2554e8
16. G. Rowell, *The Earthrise Photograph* (Australian Broadcasting Corporation)
17. de L. Broglie, "Waves and Quanta" (French: Ondes et Quanta, Presented at a Meeting of the Paris Academy of Sciences on September 10, 1923)
18. de L. Broglie, Recherches sur la théorie des quanta (Researches on the quantum theory), Thesis, Paris, 1924, Ann. de Physique **(10)** 3, 22 (1925)

D. J. Biswas, *A Beginner's Guide to Lasers and Their Applications, Part 1*,
Undergraduate Lecture Notes in Physics,
https://doi.org/10.1007/978-3-031-24330-1

19. J.D. Jackson, *Classical Electrodynamics*, 3rd edn. (Wiley, 1998)
20. Short Biography of Johann Heinrich Wilhelm Geissler; https://www.crtsite.com/Heinrich-Geissler.html
21. A.L. Shawlow, C.H. Townes, Infrared and Optical Masers. Phys. Rev. **112**, 1940 (1958)
22. C.K.N. Patel,A.K. Levine, *Gas lasers' in 'lasers: A series of advances*, vol 2 (Pub-Marcel Dekker Inc, New York)
23. T.Y. Chang, Improved uniform-field electrode profiles for TEA laser and high-voltage applications. Rev. Sci. Instrum. **44**, 405 (1973)
24. https://www.itl.nist.gov/div898/handbook/eda/section3/eda3663.htm
25. https://www.itl.nist.gov/div898/handbook/pmc/section5/pmc51.htm
26. https://www.itl.nist.gov/div898/software/dataplot/refman2/auxillar/maxpdf.htm
27. G. Pascoli, The Sagnac effect and its interpretation by Paul Langevin. Comp. Rendus Phys. **18**, 563 (2017)
28. W.E. Lamb Jr., Theory of an optical maser. Phys. Rev. A **134** (1964)
29. C.V. Raman, N.S. Nath, Proc. Ind. Acad. Sci. **2**, 406 (1935)
30. Bhagavantam, S. And Rao, B.R., Nature 161, 927 (1948)
31. W.E. Bell, Visible laser transitions in Hg. Appl. Phys. Lett. **4**, 34 (1964)
32. W.B. Bridges, Laser oscillation in singly ionized argon in the visible spectrum. Appl. Phys. Lett. **4**, 128 (1964)
33. J.V.V. Kasper, G.C. Pimentel, HCl chemical laser. Phys. Rev. Lett. **14**, 352 (1965)

Index

© The Editor(s) (if applicable) and The Author(s), under exclusive license to
Springer Nature Switzerland AG 2023
D. J. Biswas, *A Beginner's Guide to Lasers and Their Applications, Part 1*,
Undergraduate Lecture Notes in Physics,
https://doi.org/10.1007/978-3-031-24330-1

Printed in the United States
by Baker & Taylor Publisher Services